AF411330

Mechanical Properties of Advanced Multifunctional Coatings

Mechanical Properties of Advanced Multifunctional Coatings

Editors

Rong-Guang Xu
Zhitong Chen
Peijian Chen
Guangjian Peng

MDPI • Basel • Beijing • Wuhan • Barcelona • Belgrade • Manchester • Tokyo • Cluj • Tianjin

Editors
Rong-Guang Xu
The George Washington
University
Washington, DC
USA

Zhitong Chen
Chinese Academy of Sciences
Shenzhen
China

Peijian Chen
China University of Mining
and Technology
Xuzhou
China

Guangjian Peng
Zhejiang University of
Technology
Hangzhou
China

Editorial Office
MDPI
St. Alban-Anlage 66
4052 Basel, Switzerland

This is a reprint of articles from the Special Issue published online in the open access journal *Coatings* (ISSN 2079-6412) (available at: https://www.mdpi.com/journal/coatings/special_issues/adv_multifunct_coat).

For citation purposes, cite each article independently as indicated on the article page online and as indicated below:

LastName, A.A.; LastName, B.B.; LastName, C.C. Article Title. *Journal Name* **Year**, *Volume Number*, Page Range.

ISBN 978-3-0365-6670-2 (Hbk)
ISBN 978-3-0365-6671-9 (PDF)

Contents

About the Editors

Rong-Guang Xu

Rong-Guang Xu (Dr.) is currently a research fellow at Brigham and Women's Hospital and Harvard Medical School. He received his BS degree in Applied Physics from Wuhan University, Wuhan, China, and earned his Ph.D. degree at The George Washington University, Washington, D.C., USA. His research specialties include multi-scale computational physics, materials science, and mechanics, especially nanotribology, nanoconfinement, nanoindentation, computational nanomedicine, membrane fouling in water treatment, and 2D piezoelectric materials for energy harvesting. He has published about 30 journal papers, including papers in *PNAS*, *Nano Energy*, *Journal of Membrane Science*, *Langmuir*, *Journal of Chemical Physics*, *Physical Review D*, *Physics Letter B*, and *Journal of Applied Physics*.

Zhitong Chen

Zhitong Chen (Prof. Dr.) is a professor at the Institute of Biomedical and Health Engineering, Shenzhen Institute of Advanced Technology, Chinese Academy of Sciences, and holds a joint faculty affiliate appointment at the National Innovation Center for Advanced Medical Devices, China. He did his postdoctoral training in the Plasma and Space Propulsion Lab of Dr. Richard E. Wirz at the University of California, Los Angeles. He received his Ph.D. in Mechanical and Aerospace Engineering from The George Washington University, where he worked under Dr. Michael Keidar. He received his M.Sc. in Engineering Mechanics from the Institute of Mechanics, Chinese Academy of Sciences, and his B.Sc. in Mechanical Engineering from Northeast Agriculture University. His research interests include plasma physics, plasma applications, and medical devices. He has published about 50 papers, including *PNAS*, *Science Advances, and Materials Today*. He is the author of the book *Cold Atmospheric Plasma (CAP) Technology and Applications* (Springer Nature, 2021). He has applied for about 30 patents worldwide and has been granted around 10.

Peijian Chen

Peijian Chen (Prof. Dr.) is currently a full professor at the China University of Mining and Technology. He received his BS degree in Engineering Mechanics from Shandong University, Jinan, China, and earned his Ph.D. degree at LNM, the Institute of Mechanics, Chinese Academy of Sciences, Beijing, China. He worked as a visiting scholar in the Department of Mechanical Engineering at Harvard University from September 2018 to September 2019. His research specialties include surface and interface mechanics, advanced materials and structures, especially the theoretical modeling of contact mechanics, film-substrate systems, the computational simulation of 2D materials, and vdW heterostructures. He has published about 65 journal papers, including papers in *Journal of the Mechanics and Physics of Solids, International Journal of Solids and Structures, International Journal of Mechanical Sciences, Langmuir, Physical Chemistry Chemical Physics, and Mechanics of Materials. He has been granted around 10 patents worldwide.*

Guangjian Peng

Guangjian Peng (Prof. Dr.) received a Ph.D. in solid mechanics from the Institute of Mechanics, Chinese Academy of Sciences, Beijing, China, in 2013. He joined Zhejiang University of Technology, Hangzhou, China, as an assistant professor in 2013, and was made an associate professor and professor in 2015 and 2021, respectively, with the College of Mechanical Engineering. He worked as a visiting scholar in the Department of Mechanical Engineering at Johns Hopkins University from September 2017 to September 2018. He has published more than 50 research articles and is the holder of 13 patents. His current research interests include instrumented indentation techniques, the characterization of micro/nano-mechanical properties of materials, and residual stress measurement. Dr. Peng is a member of ISO/TC 164/SC 3/WG 4 and was the project leader for the development of ISO 14577-5:2022 which was published on October 31, 2022.

Preface to "Mechanical Properties of Advanced Multifunctional Coatings"

Advanced multifunctional coatings have always attracted a lot of attention due to the great and unpredicted progress in their synthesis, characterization, and properties. They have also been widely used in different fields, such as aeronautics, transportation, biomedicine, electrical and electronic equipment, etc. Mechanical properties are key to how advanced multifunctional coatings interact with external forces and environmental factors. An in-depth understanding of the mechanical properties of these coatings, however, still requires complex material modeling and characterization tools. This Special Issue aims to present the latest findings and to promote further research in the areas of mechanical behaviors of advanced multifunctional coatings, including experimental characterization and theoretical calculations. Full papers, review articles, and communications are all welcome.

Potential topics include but are not limited to the following topics:

Advanced multifunctional coatings in mechatronics;

Advanced characterization methods and tools;

Advanced coatings for preparation and applications;

Numerical simulations and computational modeling, including FEM/XFEM, MD, MC, DFT, etc.;

Theoretical studies;

Design and synthesis strategies affecting mechanical behaviors;

Industrial case studies.

Rong-Guang Xu, Zhitong Chen, Peijian Chen, and Guangjian Peng

Editors

Editorial

Special Issue: Mechanical Properties of Advanced Multifunctional Coatings

Rong-Guang Xu [1],*, Zhitong Chen [2,3],*, Peijian Chen [4],* and Guangjian Peng [5],*

[1]   Department of Mechanical and Aerospace Engineering, The George Washington University, Washington, DC 20052, USA

[2]   Center for Advanced Therapy, National Innovation Center for Advanced Medical Devices, Shenzhen 518000, China

[3]   Institute of Biomedical and Health Engineering, Shenzhen Institute of Advanced Technology, Chinese Academy of Sciences, Shenzhen 518055, China

[4]   School of Mechanics and Civil Engineering, China University of Mining and Technology, Xuzhou 221116, China

[5]   College of Mechanical Engineering, Zhejiang University of Technology, Hangzhou 310014, China

*   Correspondence: xrg117@gwmail.gwu.edu (R.-G.X.); zt.chen@nmed.org.cn (Z.C.); chenpeijian@cumt.edu.cn (P.C.); penggj@zjut.edu.cn (G.P.)

Citation: Xu, R.-G.; Chen, Z.; Chen, P.; Peng, G. Special Issue: Mechanical Properties of Advanced Multifunctional Coatings. *Coatings* **2022**, *12*, 599. https://doi.org/10.3390/coatings12050599

Received: 12 April 2022

Accepted: 25 April 2022

Published: 28 April 2022

Publisher's Note: MDPI stays neutral with regard to jurisdictional claims in published maps and institutional affiliations.

Coatings are found almost anywhere in the modern world. The typical examples are architectural wall coatings and automotive paints, which serve to both protect substrates and offer a decorative appearance [1]. In our daily life, we are surrounded by objects with coatings, such as paper books, the lenses of glasses, computer screens, CPUs inside laptops, etc. Coating technology, as one type of surface engineering, is related to the friction, wear, corrosion, abrasion, and fatigue of tools and components. They have a very wide range of applications in industries including automotive, aerospace, nuclear, military and defense, power generation, tool and die, metal forming, agriculture, and food processing. According to the industry market analysis from American Coatings Association, the total value of the global coatings market was estimated to be approximately USD 112 billion in 2014.

There has always been a lot of interest in advanced multifunctional coatings because of the tremendous and unexpected progress in their synthesis, characterization, and properties. A variety of different fields have also used them, such as aeronautics, transportation, biomedicine, equipment for electrical and electronic use, etc. Numerous excellent reviews and research articles have covered many topics in this area, such as dynamic coatings [2], thermal barrier coatings for gas-turbine engine applications [3], nanostructured coatings [4,5], all-nanoparticle thin-film coatings [6], microstructural design of hard coatings [7], anticorrosive coatings [8], advanced multifunctional coatings for vibration control of machining [9], etc. Multifunctional coatings' mechanical properties play a crucial role in their interaction with external forces and environmental factors. Yet, sophisticated characterization and modeling tools are required to gain a comprehensive understanding of the mechanical properties of these coatings.

This Special Issue aims to provide a forum for researchers to share current research findings and to promote further research into the mechanical behavior of advanced multifunctional coatings, including experimental modeling and theoretical calculations. This Special Issue eventually collects 13 original research articles on the most recent works not limited to the relevant research area of advanced multifunctional coatings, including interesting mechanical behaviors at surfaces and interfaces.

Thermal barrier coatings (TBCs) are layered materials with low thermal conductivity and high thermal stability sprayed onto metal substrates to provide thermal protection in the most demanding high-temperature environment [3,10]. As the most complex coating system, TBCs mainly consist of a ceramic top coating (TC), a metallic bonding coating (BC), and a metal substrate. The structure, mechanical properties, and failure mechanisms of TBCs would need to be understood more in order to improve them.

Thermal oxidation stress of TBCs has been successfully investigated with Cr^{3+} photoluminescence piezospectroscopy, but results for data processing lack systematization and quantitative analysis, especially regarding peak positions. Numerous spectra representing different uniaxial loadings are obtained by Lu et al. [11] through numerical experiments and calibration tests. Based on a comparison of fitting results and discussion of the generation mechanism, the Lorentzian function-rather than the more commonly utilized Psd-Voigt function—is considered to be the most relevant method for the application of Cr^{3+} photoluminescence piezospectroscopy to TBCs because it has an adequate degree of sensitivity, stability, and credibility for quantitative stress analysis.

It is one of the major causes of thermal barrier failure that residual stresses are introduced during manufacturing and service processes inside the TBC top coating. It is essential, therefore, to measure the residual stress in the top coating in a non-destructive and accurate manner in order to assess the lifetime of the TBC. Researchers have used terahertz time-domain spectroscopy (THz-TDS) to measure internal stresses in nonmetals by determining or calibrating the material's optical stress coefficient. In the work by Wang et al. [12], a THz-TDS-based calibration of the stress optical coefficient is performed for analyses of stresses in the ceramic layer of TBC. Spectra of TBC specimens are analyzed via the unimodal, multimodal, and barycenter methods of fitting THz time-domain spectra to different uniaxial compression loadings obtained with a reflection-type THz-TDS system. When compared to the two other methods, the barycenter method had a higher level of accuracy and stability. This work provides a solid foundation to employ THz-TDS for analyzing the stress inside the top coating of actual TBC structures.

A rock possesses complex structural properties in its medium space, making it one of the most basic materials for engineering coatings. However, it has still proven difficult to quantify the microstructure of rock coatings, such as the degree of connectivity and aggregation. In their first paper, an innovative new complex network theory is proposed by Gao et al. [13] to understand the 3D structure of pore networks in sandstone, which cannot be quantified by traditional methods such as Euler number and fractal dimension. Results from numerical simulations indicate that a scale-free network model is more suitable than random models for describing the pore network in sandstone. According to this research, the pore network in sandstone is uniform and its connectivity has the potential to enhance permeability. Using this method, researchers can examine crack grid distribution in tunnel coatings as well as learn more about the pore connectivity characteristics of sandstone. In their second paper from the same group [14], a dual-porosity network model of rock is proposed based on the Barabasi and Albert (BA) scale-free theory. The porosity degree distribution, average path length, throat length distribution, and other parameters of the rock model are used to quantitatively describe the connection between rock pores and throats. A trend analysis is performed for the permeability of the coating model using microscopic parameters. The dual-pore network model was shown to be capable of matching very well the structural distribution characteristics of different types of rocks, and thus can be used to describe the coating's structure very effectively. In this way, the mechanical and physicochemical properties of rock can be analyzed theoretically, providing support for the rational design of rock coatings.

The fiber-matrix interface is widely recognized as the most important determinant of composite structural stability. With an appropriate interface, load can be transferred effectively and crack propagation can be prevented. There has been considerable effort dedicated to studying how the interface affects mechanical properties of SiC–SiC composites. Through a combination of experimental measurement and theoretical analysis, Jin et al. [15] investigate the tensile properties of SiC–SiC composites reinforced by domestic Hi–Nicalon type SiC fiber at high temperature and illuminate how the interface coating affext the mechanical property variations. The experimental results show that tensile strength dramatically drops above 1200 C, and fiber-pull-out fracture mode changes to fiber-break.

Based on the subsequent FEM (finite element method) analysis, the authors claim that the variation in residual radial stress at the fiber-matrix interface coating is responsible for the reduction in tensile strength and change of fracture mode. The residual radial stress at the fiber-matrix interface coating changes from tensile to compressive once the temperature exceeds the preparation temperature for the composites, increasing the interface strength as the temperature increases.

A novel method called groove HCECB (hot casting plus explosion compression bonding) is developed by Sun et al. [16] and applied to weld 6061 aluminum alloy with Q235a steel. The experimental results demonstrate that the aluminum-steel composite plate can be directly combined by the HCECB method with no defects (such as melting layer, holes, and cracks) observed in the interface. The detailed SEM (scanning electron microscope) and EDX (Energy Dispersive X-Ray) analysis reveal that with tight bonding, and almost no melt, the composite plate has an interface consisting of flatness and microwave. According to the tensile and shear test results, the shear strength is greater than 80 MPa, which is the required bond strength for aluminum-steel composite plates.

In the experimental work by Li et al. [17], Cr-DLC (Cr-containing hydrogenated amorphous diamond-like carbon) films are synthesized by pulsed-dc (direct current) magnetron sputter with a PEM (plasma emission monitor) system using a large-size industrial Cr target. They examine how the Cr atom plasma emission intensity is related to the element concentration, the cross-sectional morphology, the deposition rate, the microstructure, mechanical properties, and tribological properties of Cr-DLC films. Various methods are employed to characterize and analyze the mechanical and tribological behaviors, including nanoindentation, scratch test instruments, and ball-on-disk reciprocating friction/wear testers. Based on the results of the experiment, the PEM system has been successfully applied to magnetron sputtering in order to deposit Cr-DLC with a more stable process.

During the last twenty years, micropitting has been particularly problematic on gear surfaces. As a surface fatigue phenomenon, it occurs in rolling and sliding contact that are operating in boundary lubrication (BL) or elastohydrodynamic lubrication (EHL) regimes. Generally, asperities on the surface are usually the source of micropitting, which starts to form in the dedendum of the driver and driven. However, Zhao et al. [18] find that for some gears with interference fit connections of their conical surface, micropitting on the pinion occurs in the addendum. Through a 3D–TCA method (GATES: Gear Analysis for Transmission Error and Stress) based on ISO/TR 15144-1:2014, this study attempts to predict the occurrence of micropitting and try to understand the key influential factors to affect micropitting location. Combined with FEM modeling, they demonstrate that instead of profile shift coefficient, the start of tip relief, and lead slope deviation, the difference of profile slope deviation between pinion and wheel may affect the micropitting location.

Nanoindentation is a widely used experimental tool to investigate mechanical responses of small volumes of materials at a micro-length scale [19]. Zhang et al. [20] apply the nanoindentation technique combined with AFM (atomic force microscope) measurement to investigate the surface morphology, average grain sizes, load-penetration depth curves, and hardness of the three SiO_2 thin films with different thicknesses (500, 1000 and 2000 nm-thick). A detailed analysis of the dependence of the hardness of the SiO_2 thin films on thickness is presented. It is found that the average intrinsic hardnesses of the 500, 1000, and 2000 nm-thick SiO_2 thin films are 11.9, 10.7, and 10.4 GPa, respectively. As the film thickness and grain size increase, the average intrinsic hardness of SiO_2 thin films decreases, in a similar manner to the Hall-Petch relationship.

Anti-reflective (AR) coatings are made up of multiple layers of metal oxides that reduce and eliminate reflection from the front and rear surfaces of ophthalmic lenses. This allows more light to pass through the lens, improving the vision of the wearer as well as decreasing the glare we see in photos. Using nanoindentation hardness as a measure of practical scratch resistance for mechanically tunable anti-reflective hard coatings, Price et al. [21] present a fundamental understanding of the relationship. It has been shown that FEM is an effective technique for analyzing the hardness of multilayer films.

Piirsoo et al. [22] carry out nanoindentation tests on the atomic layer deposited amorphous Al_2O_3-Ta_2O_5 double- and triple-layered films with thickness of 70 nm. It is found that the sequence of the oxides from surface to substrate along with the layer thickness influenced the hardness. However, the elastic modulus of amorphous Al_2O_3-Ta_2O_5 nanolaminates does not depend on the layer structure and fell between 145 and 155 GPa for all the laminates.

Computational modeling such as CFD (continuum level) and MD/MC simulations (atomistic level) can be complementary to experimental investigations. In their work, a simulation model within the framework of computational fluid dynamics (CFD) is presented by Cui et al. [23] to reveal the behavior of resin filling in the process of UV-NIL (ultraviolet nanoimprint lithography) in the moth-eye nanostructure. This model takes the boundary slip effect into account. The model allows researchers to investigate how various process parameters may influence resin filling behavior, such as resin viscosity, inlet velocity, and resin thickness. Comparison with the experimental results indicate good consistency between the simulation and the experimental results, corresponding to the ability to use the simulation model to examine the resin filling behavior of the moth-eye nanostructure in the UV-NIL process.

Xu et al. [24] perform computational simulations combined with grand canonical Monte Carlo (GCMC) and molecular dynamics (MD) to study the adsorption/desorption isotherms of argon thin films confined between commensurate/incommensurate contacts in boundary lubrication. By employing the so-called mid-density scheme to the hysteresis loops, the equilibrium structures associated with n_c and n_c + 1 layers can be obtained. Here, the critical layer number, n_c, corresponds to the monolayer thickness at which the confinement-induced liquid-like to solid-like nucleation occurs. By comparing equilibrium structures predicted by GCMC/MD simulations and those predicted by LVMD (liquid-vapor molecular dynamics) simulations, one can determine the correct values of chemical potentials. Simulations using GCMC/MD can be used to construct equilibrium structures at varying thicknesses that may be suitable as an initial stage for studies using advanced sampling methods.

Funding: There is no funding support.

Conflicts of Interest: The authors declare no conflict of interest.

References

1. Schwalm, R. *UV Coatings: Basics, Recent Developments and New Applications*; Elsevier: Amsterdam, The Netherlands, 2006.
2. Righetti, P.G.; Gelfi, C.; Verzola, B.; Castelletti, L. The state of the art of dynamic coatings. *Electrophoresis* **2001**, *22*, 603–611. [CrossRef]
3. Padture, N.P.; Gell, M.; Jordan, E.H. Thermal barrier coatings for gas-turbine engine applications. *Science* **2002**, *296*, 280–284. [CrossRef]
4. He, J.; Schoenung, J.M. Nanostructured coatings. *Mat. Sci. Eng. A* **2002**, *336*, 274–319. [CrossRef]
5. Tjong, S.C.; Chen, H. Nanocrystalline materials and coatings. *Mater. Sci. Eng. R Rep.* **2004**, *45*, 1–88. [CrossRef]
6. Lee, D.; Rubner, M.F.; Cohen, R.E. All-nanoparticle thin-film coatings. *Nano Lett.* **2006**, *6*, 2305–2312. [CrossRef]
7. Mayrhofer, P.H.; Mitterer, C.; Hultman, L.; Clemens, H. Microstructural design of hard coatings. *Prog. Mater. Sci.* **2006**, *51*, 1032–1114. [CrossRef]
8. Sørensen, P.A.; Kiil, S.; Dam-Johansen, K.; Weinell, C.E. Anticorrosive coatings: A review. *J. Coat. Technol. Res.* **2009**, *6*, 135–176. [CrossRef]
9. Rashid, M.-U.; Guo, S.; Adane, T.F.; Nicolescu, C.M. Advanced multi-functional coatings for vibration control of machining. *J. Mach. Eng.* **2020**, *20*, 5–23. [CrossRef]
10. Vaßen, R.; Jarligo, M.O.; Steinke, T.; Mack, D.E.; Stöver, D. Overview on advanced thermal barrier coatings. *Surf. Coat. Technol.* **2010**, *205*, 938–942. [CrossRef]
11. Lu, N.; Zhang, Y.; Qiu, W. Comparison and selection of data processing methods for the application of Cr^{3+} photoluminescence piezospectroscopy to thermal barrier coatings. *Coatings* **2021**, *11*, 181. [CrossRef]
12. Wang, Z.; Zhang, Y.; Lu, N.; Wang, Z.; Qiu, W. Measurement of stress optic coefficient for thermal barrier coating based on terahertz time-domain spectrum. *Coatings* **2021**, *11*, 1265. [CrossRef]
13. Gao, F.; Hu, Y.; Liu, G.; Yang, Y. Experiment study on topological characteristics of sandstone coating by micro CT. *Coatings* **2020**, *10*, 1143. [CrossRef]

14. Ye, D.; Liu, G.; Gao, F.; Zhu, X.; Hu, Y. A study on the structure of rock engineering coatings based on complex network theory. *Coatings* **2020**, *10*, 1152. [CrossRef]

15. Jin, E.; Sun, W.; Liu, H.; Wu, K.; Ma, D.; Sun, X.; Feng, Z.; Li, J.; Yuan, Z. Effect of interface coating on high temperature mechanical properties of SiC–SiC composite using domestic Hi–nicalon type SiC fibers. *Coatings* **2020**, *10*, 477. [CrossRef]

16. Sun, Y.-L.; Ma, H.-H.; Yang, M.; Shen, Z.-W.; Luo, N.; Wang, L.-Q. Research on interface characteristics and mechanical properties of 6061 Al alloy and Q235a steel by hot melt-explosive compression bonding. *Coatings* **2020**, *10*, 1031. [CrossRef]

17. Li, G.; Xu, Y.; Xia, Y. Effect of Cr atom plasma emission intensity on the characteristics of Cr-DLC films deposited by pulsed-DC magnetron sputtering. *Coatings* **2020**, *10*, 608. [CrossRef]

18. Zhao, L.; Shao, Y.; Du, M.; Yang, Y.; Bian, J. The influence of pulling up on micropitting location for gears with interference fit connections of their conical surface. *Coatings* **2020**, *10*, 1224. [CrossRef]

19. Schuh, C.A. Nanoindentation studies of materials. *Mater. Today* **2006**, *9*, 32–40. [CrossRef]

20. Zhang, W.; Li, J.; Xing, Y.; Nie, X.; Lang, F.; Yang, S.; Hou, X.; Zhao, C. Experimental study on the thickness-dependent hardness of SiO$_2$ thin films using nanoindentation. *Coatings* **2020**, *11*, 23. [CrossRef]

21. Price, J.J.; Xu, T.; Zhang, B.; Lin, L.; Koch, K.W.; Null, E.L.; Reiman, K.B.; Paulson, C.A.; Kim, C.-G.; Oh, S.-Y.; et al. Nanoindentation hardness and practical scratch resistance in mechanically tunable anti-reflection coatings. *Coatings* **2021**, *11*, 213. [CrossRef]

22. Piirsoo, H.-M.; Jõgiaas, T.; Ritslaid, P.; Kukli, K.; Tamm, A. Influence to hardness of alternating sequence of atomic layer deposited harder alumina and softer tantala nanolaminates. *Coatings* **2022**, *12*, 404. [CrossRef]

23. Cui, Y.; Wang, X.; Zhang, C.; Wang, J.; Shi, Z. Numerical investigation for the resin filling behavior during ultraviolet nanoimprint lithography of subwavelength moth-eye nanostructure. *Coatings* **2021**, *11*, 799. [CrossRef]

24. Xu, R.-G.; Rao, Q.; Xiang, Y.; Bian, M.; Leng, Y. Computational simulations of nanoconfined argon film through adsorption–desorption in a uniform slit pore. *Coatings* **2021**, *11*, 177. [CrossRef]

Article

Comparison and Selection of Data Processing Methods for the Application of Cr³⁺ Photoluminescence Piezospectroscopy to Thermal Barrier Coatings

Ning Lu [1], Yanheng Zhang [1] and Wei Qiu [1,2,*]

[1] Department of Mechanics, Tianjin University, Tianjin 300354, China; sugariss@tju.edu.cn (N.L.); tjuzhangyh@tju.edu.cn (Y.Z.)
[2] Tianjin Key Laboratory of Modern Engineering Mechanics, Tianjin 300354, China
*   Correspondence: qiuwei@tju.edu.cn

Abstract: Thermal barrier coatings (TBCs) are an indispensable part of the blades used in aeroengines. Under a high-temperature service environment, the thermal oxidation stress at the interface is the main cause of thermal barrier failure. Cr^{3+} photoluminescence piezospectroscopy has been successfully used to analyze the thermal oxidation stress of TBCs, but systematic and quantitative analysis results for use in data processing are still lacking, especially with respect to the identification of peak positions. The processing methods used to fit spectral data were studied in this work to accurately characterize TBC thermal oxidation stress using Cr^{3+} photoluminescence spectroscopy. Both physical and numerical experiments were carried out, where Cr^{3+} photoluminescence spectra were detected from alumina ceramic samples under step-by-step uniaxial loading, and the simulated spectra were numerically deduced from the measured spectral data. Then, the peak shifts were obtained by fitting all spectral data by using Lorentzian, Gaussian and Psd-Voigt functions. By comparing the fitting results and then discussing the generation mechanism, the Lorentzian function—not the Psd-Voigt function that is most widely utilized—was regarded as the most applicable method for the application of Cr^{3+} photoluminescence piezospectroscopy to TBCs because of its sufficient sensitivity, stability and confidence for quantitative stress analysis.

Keywords: thermal barrier coatings; thermal oxidation stress; Cr^{3+} photoluminescence piezospectroscopy; data processing method; fitting function; Lorentzian functions; spectromechanics

Citation: Lu, N.; Zhang, Y.; Qiu, W. Comparison and Selection of Data Processing Methods for the Application of Cr³⁺ Photoluminescence Piezospectroscopy to Thermal Barrier Coatings. *Coatings* **2021**, *11*, 181. https://doi.org/10.3390/coatings11020181

Academic Editor: Cecilia Bartuli
Received: 30 December 2020
Accepted: 1 February 2021
Published: 4 February 2021

Publisher's Note: MDPI stays neutral with regard to jurisdictional claims in published maps and institutional affiliations.

1. Introduction

Thermal barrier coatings (TBCs) have been developed to effectively protect gas turbines and aeroengines by covering the surface of blade substrates with ceramic materials [1–3]. However, during high-temperature service, thermally grown oxide (TGO) will grow at the interface between the ceramic layer and bonding layer [4–7], which causes the generation and evolution of thermal oxidation stress, leading to interfacial and internal cracks and spalling and even the failure of TBCs [8–12]. Therefore, theoretical modeling, numerical simulation and experimental measurement are all indispensable means of analyzing the state, magnitude, distribution and evolution of the stress induced by TGO, which are essential for scientific research and engineering investigations of TBCs. As an optical measurement method, photoluminescence piezospectroscopy can be applied to investigate a mechanical system non-destructively [13].

α-Al_2O_3 is the main component of TGO; it is formed from the chemical reaction between Al in the bonding layer and oxygen penetrating through the TBC layer during high-temperature service. Because traces of Cr^{3+} inside the bonding layer are doped into α-Al_2O_3 during its growth, the noncontact Cr^{3+} photoluminescence piezospectroscopy technique is regarded as an effective means of measuring thermal oxidation stress in TGO [14–17]. In 1993, Ma and Clarke found a quantitative relationship between the stress and the peak position of the Cr^{3+} photoluminescence spectra of some monocrystalline and polycrystalline materials, constructing the theoretical foundation of fluorescence piezospectroscopy in ceramics [18]. By using a fluorescence piezospectroscopy method, the stress of TGO in TBCs produced by electron beam physical vapor deposition (EB-PVD) was determined by Christensen et al. [19], in 1996. In 2001, Nychka et al. [20] further studied the evolution of thermal cycling stress in TGO. Schlichting et al. [21] discovered the relationship between the stress and thickness of the ceramic layer by measuring the average stress in TGO. In 2019, Shen et al. [22] measured the compressive stress of TGO under thermal shock cycles and found that the evolution of the TGO microstructure, cracks, residual stress and element depletion appeared to be relevant to the durability and degradation of TBCs. In 2020, Jiang et al. [23] verified that the residual stress evolution of TGO for air plasma spraying TBCs can be effectively used to indicate interfacial delamination.

The identification of the peak position is one of the key procedures used with Cr^{3+} photoluminescence spectroscopy to characterize the internal stress of TGO. In most published works, the pseudo-Voigt (Psd-Voigt) function has been most widely applied to fit Cr^{3+} photoluminescence spectra, and the stress in TGO was generally at a magnitude of hundreds of MPa and even GPa. For example, the Psd-Voigt function was used to fit Cr^{3+} fluorescence spectra for stress analyses by Clarke et al. [24] and revealed that the compressive stress in the TGO was approximately 3 to 4 GPa during oxidation. Seculk et al. [25] used a Lorentzian–Gaussian (viz. Psd-Voigt) curve to fit the peaks of the Cr^{3+} luminescence spectra, and the measured stress ranged from 1 GPa in tension to 5 GPa in compression. The evolution of thermal oxidation stress in TGO was studied by Manero et al. [26] using the Psd-Voigt fitting function, and the stress detected ranged from approximately 2 to 4 GPa throughout the loading history, with a standard deviation of approximately 78 MPa under the same loading conditions. Jiang et al. [23] employed a Lorentzian curve algorithm to fit the spectrum obtained from TGO and determined the peak position, and the average stress gradually increased from 1.76 GPa until a peak value of approximately 3.45 GPa was reached. Shen et al. [22] discovered that the compressive stress of TGO with a LZC/YSZ top coating reached the highest value of 2.56 GPa at 1000 thermal cycles, but the data processing method was not mentioned. In addition, the interfacial crack propagation behavior induced by thermal oxidation stress is very complicated because of the undulating TGO morphology [27], and the microcracks always nucleate at local undulations, which continuously causes a slight relief in the stress in the TGO [28]. Therefore, only high-resolution stress characterization can sufficiently explain the dynamic evolution mechanism of TBC delamination, which has been effectively presented by the latest numerical model [29]. However, among all the published works, the residual stress measured in TGO varied by at least tens of MPa, which ineluctably made accurately determining the evolution tendency difficult because of the low resolution of the stress measurement [22,23].

Because blades must be in long-term and stable service in extremely hot and humid environments, it is necessary to develop physical-chemical-mechanical models to quantitively describe the thermal oxidation stress in TBCs and nondestructive methods to reliably evaluate the structural safety of the TBCs, with both measures requiring accurate in situ stress characterization with precision in the order of ten MPa. However, as mentioned above, the resolution and confidence of the stress characterization based on Cr^{3+} photoluminescence piezospectroscopy have generally been in the order of hundreds of MPa.

Cr^{3+} photoluminescence piezospectroscopy is not the only spectromechanical method. Other techniques, such as cathodoluminescence spectroscopy [30,31], Eu^{3+} photoluminescence piezospectroscopy [11,32], and micro-Raman spectroscopy [33–38], have been broadly applied to the stress/strain analyses of different materials and structures. Fitting functions, such as Lorentzian, Gaussian, and Psd-Voigt functions, have been used to identify the positions and intensities of the characteristic peaks in the spectra detected from samples with external or internal stress. However, there is still a lack of specific studies on the effectiveness and applicability of data processing methods for spectromechanical investigations, especially the high-resolution stress analysis of TBCs using Cr^{3+} photoluminescence piezospectroscopy.

In this work, several fitting functions and methods were compared by applying them to identify the peak positions of Cr^{3+} photoluminescence spectra. The fitted spectra were obtained through both numerical experiments and calibration tests under uniaxial loading. Subsequently, the increments of the peak positions corresponding to different stresses were characterized by different data processing methods, including Lorentzian, Gaussian, and Psd-Voigt functions. Then, the piezo-spectroscopic coefficients were linearly fitted based on the correlation between the peak shift and uniaxial stress. The sensitivity and reliability of the different fitting functions were verified and compared through experimental results to obtain the best one for Cr^{3+} photoluminescence piezospectroscopy.

2. Materials and Experiments

The main component of TGO is α-Al_2O_3, which is formed from the chemical reaction during high-temperature service, and its thickness is lower than 10 μm. Therefore, it is difficult to extract TGO from TBCs separately. However, Cr^{3+} photoluminescence can be obtained from the traces of Cr^{3+} doped in polycrystalline alumina ceramics, which have been applied to the calibration of piezo-spectroscopic coefficients by Ma et al. [18]. An alumina ceramic specimen was produced and used to obtain Cr^{3+} fluorescence spectra under a uniaxial step stress state in this paper. This specimen was prepared by high-temperature sintering and then cut into several $4 \times 4 \times 10$ mm^3 samples.

As shown in Figure 1a, the experimental system of the calibration test mainly consisted of three devices, including a self-built micro-Raman/PL system using a 500 mm spectroscope (Shamrock 500i, Andor, Belfast, UK), a uniaxial loading device (Microtest 5000, Deben, Suffolk, UK) with a load capacity of 5000 N and an electric 2D-stage (RC201ZA100 × 100, Beijing Ruicheng, Beijing, China). In the spectroscopic system, a 532 nm laser was used to induce Cr^{3+} photoluminescence in the alumina ceramic samples. The spectral resolution was approximately 0.01 nm, and an 1800 L/mm grating was used. A 20× objective lens (378-810-3, Mitutoyo, Kawasaki, Japan) was chosen to achieve a sampling spot with a diameter of ~10 μm. Before the calibration test, an alumina ceramic sample was fixed on the loading device through a compressive preload of 200 N. Then, the sample was loaded step-by-step by compressive forces to 3000 N with a step length of 200 N (according to 12.5 MPa per step for this sample). At each loading step, luminescence spectra were recorded from 20 different sampling spots on the top surface of the sample. Figure 1b shows a typical Cr^{3+} fluorescence spectrum with 2 characteristic peaks, namely, R_1 and R_2.

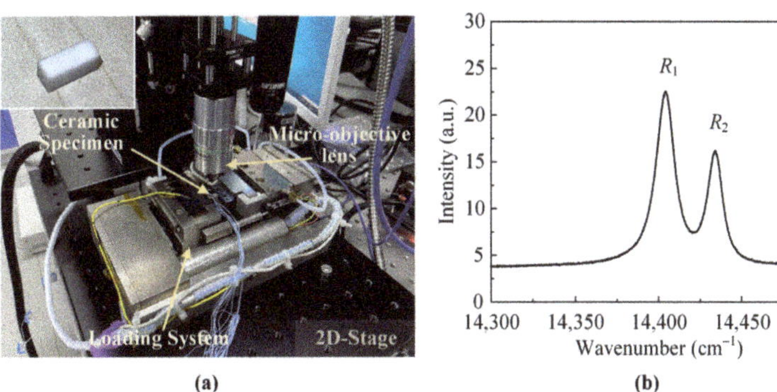

Figure 1. (a) The experimental system of the calibration test; (b) a typical Cr^{3+} fluorescence spectrum.

To analyze the applicability of each processing method for accurately identifying the wavenumber increment of the Cr^{3+} luminescence peak caused by stress, numerical experiments were carried out based on actually measured spectra (shown in Figure 1b) from an unloaded specimen. The procedure used for the numerical experiments is shown in Figure 2. First, a smooth fitting curve (shown in Figure 3a) that had an extreme goodness-of-fit (GOF) over 0.9995 was obtained through multipeak Lorentzian function fitting via the OriginPro 2016 software (OriginLab, Northampton, MA, USA). The GOF in this work was based on the substitution sum of squares of the residuals between the fitted curve and the actual spectrum data. The noise distribution of the actual spectrum relative to the fitting curve is given in Figure 3b. The random distribution follows a normal distribution, whose center is −0.0002 and FWHM (full width at half maximum) is 0.0942. The data density of the fitted curve was increased to 30-fold that of the experimental data obtained by interpolation.

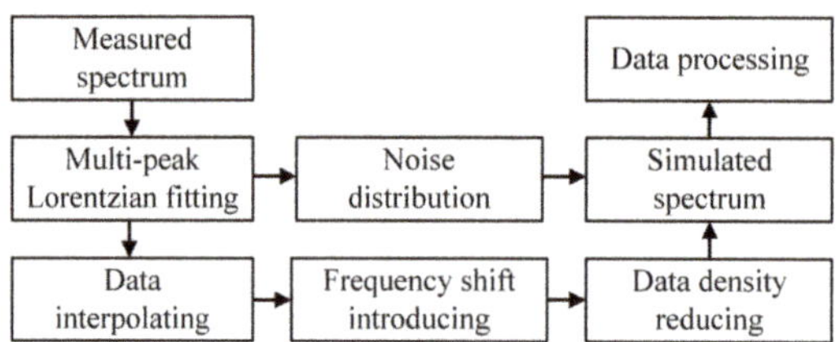

Figure 2. Procedure used for the numerical experiments.

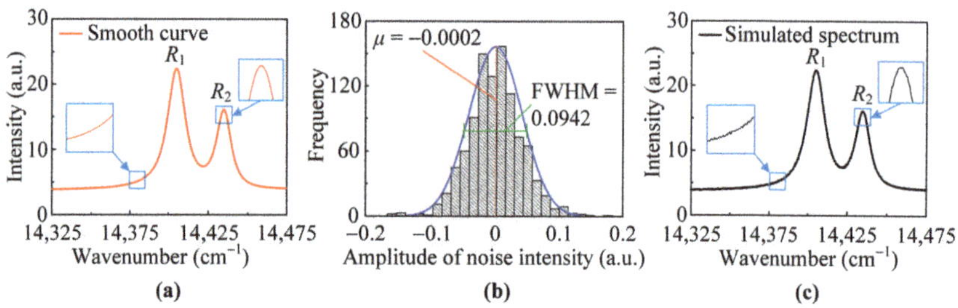

Figure 3. (**a**) Smooth curve obtained by multipeak fitting; (**b**) noise distribution; (**c**) simulated spectrum with introduced noise.

After that, some overall shifts in the wavenumbers were introduced into the density-increased curve by adding certain increments to the horizontal ordinate data, which was aimed to simulate the effect of stress on the spectral curve. Both the wavenumber-stress coefficients of R_1 and R_2 were regarded as −1.45 cm^{-1}/GPa, and the increments were set to a fixed step length of 0.007 cm^{-1} at the beginning, which corresponded to 5 MPa. Then, the incremental step was increased step-by-step.

Finally, the data density of each wavenumber-shifted curve was reduced to a degree consistent with the actually measured spectrum, and random errors at the same level as the error of the actually measured spectrum were introduced using a normal distribution (shown in Figure 3b), achieving the spectral data of the numerical experiments, as shown in Figure 3c.

3. Models and Methods

3.1. The Theoretical Model of Cr^{3+} Photoluminescence Piezospectroscopy

The basis for using piezo-spectroscopic methods for the measurement of stress in crystalline materials is that an applied stress strains the lattice and alters the energy of transitions between electronic or vibrational states [39,40]. Cr^{3+} ions exist in α-Al$_2$O$_3$ corundum in trace impurity form. Cr^{3+} ions can be used to substitute for some Al^{3+} ions, occupying the center of normal octahedral ion sites, since they are similar in ionic radius. When α-Al$_2$O$_3$ corundum is subjected to pressure, there is an obvious wavenumber increment of Cr^{3+} fluorescence characteristic peak with suitable excitation.

The relationship between an observed shift in a fluorescence line and the state of stress was first described phenomenologically by Grabner [41], who proposed that the change in frequency, $\Delta\nu$, of the fluorescence line can be expressed by the tensorial equation

$$\Delta\nu = \Pi_{ij}\sigma_{ij}{}^* = \begin{bmatrix} \Pi_{11} & \Pi_{12} & \Pi_{13} \\ \Pi_{21} & \Pi_{22} & \Pi_{23} \\ \Pi_{31} & \Pi_{32} & \Pi_{33} \end{bmatrix} \begin{bmatrix} \sigma_{11}{}^* & \sigma_{12}{}^* & \sigma_{13}{}^* \\ \sigma_{21}{}^* & \sigma_{22}{}^* & \sigma_{21}{}^* \\ \sigma_{31}{}^* & \sigma_{32}{}^* & \sigma_{33}{}^* \end{bmatrix} \tag{1}$$

where Π_{ij} denotes the component of the wavenumber-stress coefficient tensor, $\sigma_{ij}{}^*$ is the stress component, and the subscripts i and j represent the crystallographic directions. The Π_{ij} matrix can be simplified by considering the point symmetry of Cr^{3+} ions [39]:

$$\Pi_{ij} = \begin{bmatrix} \Pi_{11} & 0 & 0 \\ 0 & \Pi_{22} & 0 \\ 0 & 0 & \Pi_{33} \end{bmatrix} \tag{2}$$

Equation (1) reduces to

$$\overline{\Delta\nu} = \Pi_{11}\sigma_{11}{}^* + \Pi_{22}\sigma_{22}{}^* + \Pi_{33}\sigma_{33}{}^* \tag{3}$$

The main composition of the TGO is polycrystalline α-Al$_2$O$_3$, which satisfies the assumption of macroscopic isotropy. There are enough grains with random crystal orientations in the collecting area during laser excitation in the PL measurement. The spatial orientation of the piezo-spectroscopic relationship can be expressed by the Euler angle transformation matrix, and then, the effect of off-diagonal elements Π_{ij} $(i \neq j)$ and $\sigma_{ij}{}^*$ $(i \neq j)$ vanishes when integrating over the whole space. Therefore, Equation (3) reduces to:

$$\overline{\Delta\nu} = \frac{1}{3}(\Pi_{11} + \Pi_{22} + \Pi_{33})(\sigma_{11} + \sigma_{22} + \sigma_{33}) \tag{4}$$

Furthermore, when uniaxial stress, σ_U, is applied, the frequency shift, $\Delta\nu$, can be expressed as follows:

$$\overline{\Delta\nu} = \Pi_U\sigma_U \tag{5}$$

where Π_U is the piezo-spectroscopic coefficient under the uniaxial stress state.

3.2. Data Processing Methods

Previous studies have shown that the fluorescence peaks at 14,402 and 14,432 cm^{-1} (shown in Figure 1b) are highly sensitive to stress [42–49]. Lorentzian, Gaussian and Psd-Voigt functions were mostly used to fit the characteristic peaks or bands in the experimental molecular and atomic spectral data, including those obtained by Cr^{3+} photoluminescence [24–26]. In this work, all the luminescence spectra of both actual and numerical experiments were fitted by using Lorentzian, Gaussian and Psd-Voigt functions. The spectral fitting was carried out based on Levenberg–Marquardt algorithm via the software OriginPro 2016.

The line shape accords with the Lorentzian function for the homogeneous broadening of the spectrum caused by spontaneous emission and collision. The expression of the Lorentzian function is as follows:

$$y = y_0 + \frac{2A}{\pi} \times \frac{w}{4(x - x_c)^2 + w^2} \tag{6}$$

where y_0 is the relative base value of the peak, A is the area of this peak, w is its FWHM, and x_c is the peak position.

The line shape accords with the Gaussian function for the inhomogeneous broadening of spectrum caused by the velocity distribution of irradiance particles (viz. Doppler effect). The expression of the Gaussian function is as follows:

$$y = y_0 + \frac{A e^{\frac{-4\ln 2}{w^2}(x-x_c)^2}}{w\sqrt{\frac{\pi}{4\ln 2}}} \tag{7}$$

where y_0, A, w and x_c have the same meaning as they do in Equation (6).

The abovementioned homogeneous broadening caused by collision and inhomogeneous broadening caused by the Doppler effect are the main two types of broadening, but spectral lines are always characterized by a variety of broadening mechanisms in general. When considering these above two kinds of broadening mechanisms simultaneously, the Psd-Voigt function, which is the linear superposition of a Lorentzian function and a Gaussian function, can be obtained. The expression is as follows:

$$y = y_0 + A\left[m_u \frac{2}{\pi} \times \frac{w_L}{4(x-x_c)^2 + w_L^2} + (1-m_u)\frac{\sqrt{4\ln 2}}{\sqrt{\pi}w_G}e^{-\frac{4\ln 2}{w_G^2}(x-x_c)^2} \right] \tag{8}$$

where w_L is the FWHM of the Lorentzian curve, w_G is the FWHM of the Gaussian curve, x_c is the peak position, and m_u is the weight of the Lorentzian function.

4. Results and Discussion

The fitting curves of a typical Cr^{3+} photoluminescence spectrum obtained using Psd-Voigt, Gaussian and Lorentzian functions are shown in Figure 4a–c, respectively. The GOFs of different functions for both the measured and simulated spectra are listed in Table 1. Fitting using the Psd-Voigt function can obtain a visually better fitting effect (shown in Figure 4) and a larger value of the GOF (listed in Table 1) than the other two functions. In contrast, the fitting result obtained using the Gaussian function shows the largest shape difference and the lowest GOF.

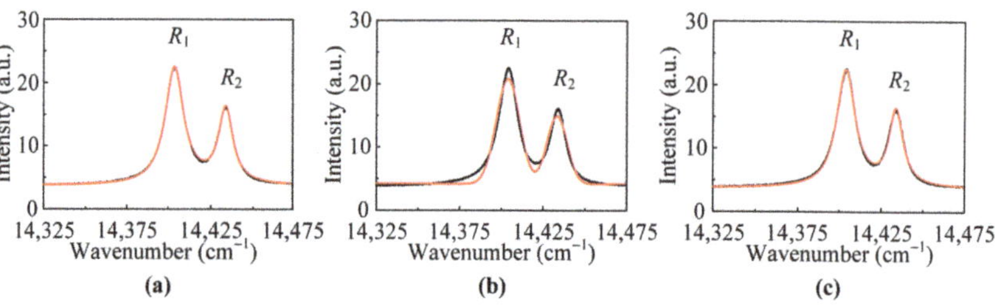

Figure 4. Typical Cr^{3+} photoluminescence spectra fit using (a) Psd-Voigt, (b) Gaussian and (c) Lorentzian functions.

Table 1. GOF values of different functions used to fit Cr^{3+} fluorescence spectra.

Fitting Function		Psd-Voigt	Gaussian	Lorentzian
GOF	Simulated data	0.9994	0.9808	0.9986
	Measured data	0.9991	0.9804	0.9986

4.1. Results of the Numerical Experiments

All the simulated spectra were fitted by using Psd-Voigt, Gaussian and Lorentzian functions and the simulated spectral data corresponding to specific stress values were exactly the same. In addition, the wavenumber increments, viz., the shifts of the R_1 and

R_2 peaks, were obtained as shown in Figures 5–7. Based on the fitting results, piezo-spectroscopic coefficients Π_U were obtained and are listed in Table 2.

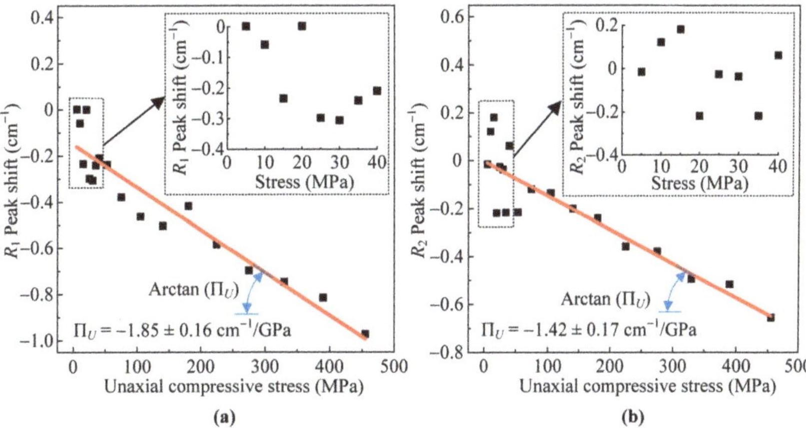

Figure 5. The relation between the peak shift and stress fitted by the Psd-Voigt function: (**a**) peak R_1; (**b**) peak R_2.

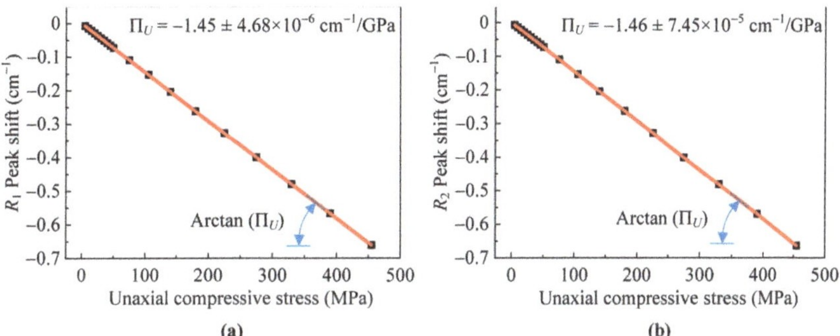

Figure 6. The relation between the peak shift and stress fitted by the Gaussian function: (**a**) peak R_1; (**b**) peak R_2.

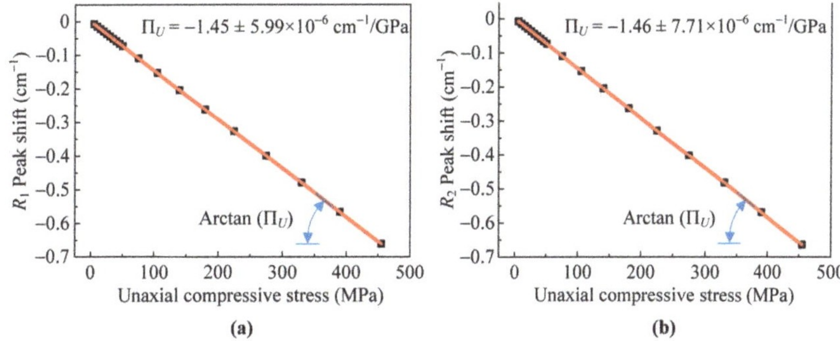

Figure 7. The relation between the peak shift and stress fitted by the Lorentzian function: (**a**) peak R_1; (**b**) peak R_2.

Table 2. Piezo-spectroscopic coefficients Π_U in numerical experiments based on the results obtained by using different fitting functions.

Π_U	R_1 (cm^{-1}/GPa)	R^2	R_2 (cm^{-1}/GPa)	R^2
Preset value	-1.45	-	-1.45	-
Psd-Voigt	-1.85 ± 0.16	0.8964	-1.42 ± 0.17	0.8118
Gaussian	$-1.45 \pm 4.68 \times 10^{-6}$	1	$-1.46 \pm 7.45 \times 10^{-5}$	1
Lorentzian	$-1.45 \pm 5.99 \times 10^{-6}$	1	$-1.46 \pm 7.71 \times 10^{-6}$	1

Figure 5 gives the results of fitting peak shifts under different uniaxial loadings (equal step length and then increasing step length) by using the Psd-Voigt function. On the whole, the peak shift, whether R_1 or R_2, maintained a linear relationship with the compressive stress, and the slope of this linear relationship is called the piezo-spectroscopic coefficient, denoted as Π_U. However, such a linear relationship did not exist at the small loading step (corresponding to a ~5 MPa stress pre-step), as shown in the upper right subfigures of Figure 5a,b, where the peak shifts changed with a seemingly random increment. This means that the Psd-Voigt function was not sensitive enough to small stress when fitting the Cr^{3+} photoluminescence spectrum. Moreover, the peak shift began to change monotonically in Figure 5 under the condition of larger loading steps, but the data still obviously fluctuated, and the results were not correct. Specifically, the linear correlation coefficients (R_2) between the peak shift and stress were only 0.8964 for R_1 and 0.8118 for R_2. In addition, Π_U is -1.85 ± 0.16 cm^{-1}/GPa for R_1 and -1.42 ± 0.17 cm^{-1}/GPa for R_2, both of which are quite different from the preset value of -1.45 cm^{-1}/GPa.

Figure 6 gives the fitting results obtained by using the Gaussian function, which shows that the shifts of both the R_1 peak and R_2 peak decreased linearly ($R^2 = 1$) with increasing uniaxial stress. Π_U was $-1.45 \pm (4.68 \times 10^{-6})$ cm^{-1}/GPa for R_1 and $-1.46 \pm (7.45 \times 10^{-5})$ cm^{-1}/GPa for R_2, each of which was nearly equal to the preset value -1.45 cm^{-1}/GPa. A similar phenomenon occurred when using Lorentzian functions. The peak shift, whether R_1 or R_2, maintained a faultlessly linear relationship with the compressive stress. Π_U was $-1.45 \pm (5.99 \times 10^{-6})$ cm^{-1}/GPa for R_1 and $-1.46 \pm (7.71 \times 10^{-6})$ cm^{-1}/GPa for R_2.

4.2. Results of the Calibration Experiments

All the experimental spectra were fitted by using Psd-Voigt, Gaussian and Lorentzian functions. The wavenumber increments were obtained, as shown in Figures 8–10. Based on the fitting results, the piezo-spectroscopic coefficients Π_U were obtained and are listed in Table 3. By comparing these results using the three fitting functions, it can be seen that the peak shifts obtained by the Psd-Voigt function had the largest fluctuation; that is, they had the poorest stability. Specifically, the peak shifts of the different sampling spots under the same loading changed over large ranges, with averages of 0.21 cm^{-1} for the R_1 peak and 0.13 cm^{-1} for the R_2 peak. The peak shifts changed with external loadings in the linear relationship of low linear correlation coefficients, approximately 0.97 for the R_1 peak and 0.97 for the R_2 peak.

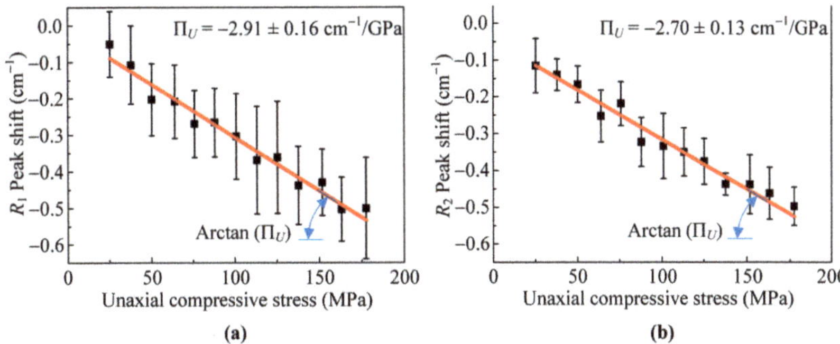

Figure 8. The relation between the peak shift and stress fitted by the Psd-Voigt function: (**a**) peak R_1; (**b**) peak R_2.

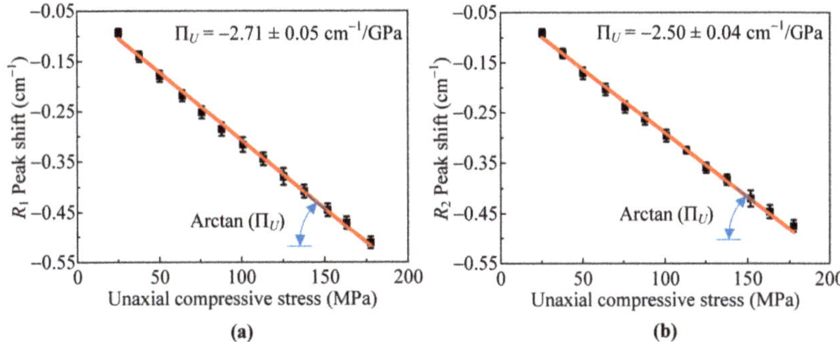

Figure 9. The relation between the peak shift and stress fitted by the Gaussian function: (**a**) peak R_1; (**b**) peak R_2.

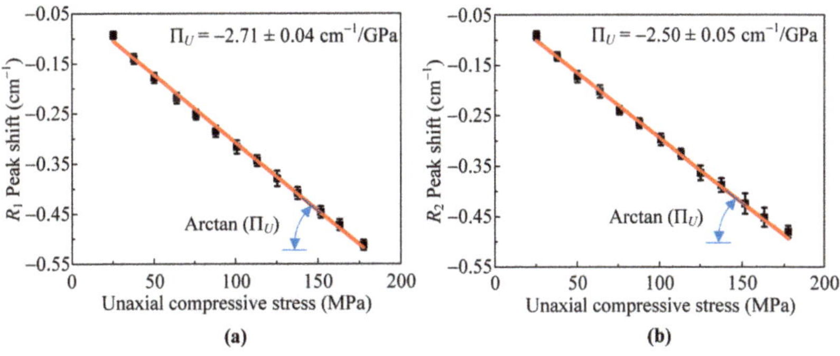

Figure 10. The relation between the peak shift and stress fitted by the Lorentzian function: (**a**) peak R_1; (**b**) peak R_2.

Table 3. Piezo-spectroscopic coefficients Π_U in numerical experiments based on the results obtained by using different fitting functions.

Π_U	R_1 (cm^{-1}/GPa)	R^2	R_2 (cm^{-1}/GPa)	R^2
Ma, 1993 [18]	-2.54	-	-2.53	-
Psd-Voigt	-2.91 ± 0.16	0.9684	-2.70 ± 0.13	0.9732
Gaussian	-2.71 ± 0.05	0.9969	-2.50 ± 0.04	0.9966
Lorentzian	-2.71 ± 0.04	0.9970	-2.50 ± 0.05	0.9961

In contrast, the peak shifts fitted by both Gaussian (shown in Figure 9) and Lorentzian (shown in Figure 10) functions decreased linearly with increasing uniaxial stress for both R_1 and R_2. The piezo-spectroscopic coefficient Π_U obtained by the Gaussian function was almost equal to that obtained by the Lorentzian function, -2.71 cm^{-1}/GPa for R_1 and -2.50 cm^{-1}/GPa for R_2. Notably, there were few differences between the Π_U results of this work and those in a published work [18], where Π_U was -2.54 cm^{-1}/GPa for R_1 and -2.53 cm^{-1}/GPa for R_2.

4.3. Discussion

The fitting results above show that, even though it had a visually better fitting effect and higher GOF, the Psd-Voigt function is not suitable for the stress analysis of Cr^{3+} photoluminescence piezospectroscopy, owing to its low sensitivity to small stress changes, and low stability and confidence of peak shifts. Particularly, it did not obviously identify a stress change with a range of 40 MPa, as shown in the upper right subfigures of Figure 5. In contrast, both Gaussian and Lorentzian functions accurately identified the peak shift caused by small stresses and stably described the linear relationship between stress and peak shift. By comparison, the Lorentzian function always achieved a higher GOF, as well as a better visual similarity, than the Gaussian function. Hence, by comparing and contrasting the fitting effect, the Lorentzian function is proven to be most applicable to fit Cr^{3+} photoluminescence spectra for experimental stress analysis.

The applicability of any fitting function to spectral analysis is determined by the generation mechanism of the spectrum. Generally, a characteristic peak corresponding to a single vibration/excitation mode in the Raman/fluorescence spectrum of a crystal, especially that of a monocrystalline material, always presents as a perfect Lorentzian shape, such as the Raman G peak of graphene [50], the first-order Raman peak of monocrystalline silicon [36] and the fluorescence peak of ruby crystals. For polycrystalline materials or eutectic substances with some hybrid structures or multioverlay vibration/excitation modes, the characteristic peaks in their Raman or fluorescence spectra show a Gaussian shape. In the majority of cases, most Raman or fluorescence peaks reflect both Lorentzian and Gaussian characteristics, which are, in fact, the convolution of the two; that is, Voigt shapes. However, it is difficult to realize curve fitting using a true Voigt function for the discrete sequence of spectral data. As an alternative, the Psd-Voigt function is widely used. A Psd-Voigt function is the linear combination of a Lorentzian function and Gaussian function with a certain weight ratio.

The fitting of Raman or fluorescence spectra using a Psd-Voigt function usually obtains a relatively optimal GOF. However, the unilateral pursuit of such a fitting process is to achieve a GOF that is as high as possible by adjusting the Lorentzian–Gaussian weight ratio, which has no physical meaning. Hence, the weight ratios obtained from fitting the spectra of the same sample with different loadings or different positions are uncertain, inconsistent and irregular. Due to the loss of a unified standard for mutual comparison, it is difficult to reflect the small wavenumber changes caused by stress in the fitting results, which makes the Psd-Voigt function insensitive and inaccurate for use in stress analysis.

In fact, the Psd-Voigt function was the most widely utilized function in published works on the application of Cr^{3+} photoluminescence piezospectroscopy [24,26]. In the existing works, the measured thermal oxidation stress was always in the magnitude of hundreds of MPa or even thousands of MPa, showing that the Psd-Voigt function is not sensitive to the stress in the magnitude of tens of MPa. However, with the improvement of engineering requirements for advanced thermal barrier coating technology, accurate calibration of the piezo-spectroscopic coefficient and creditable characterization of residual/processing stress require that the stress resolution of piezo-spectroscopic analysis reach a magnitude of 10 MPa.

The Lorentzian function is more applicable to fit Cr^{3+} photoluminescence piezo spectra. A typical TBC structure is composed of a nickel alloy blade as the substrate, a bonding layer (NiCoCrAlY) and a TBC layer on the surface. During its service under high temperature (at least 1400 K), thermal oxidation stress is generated and evolves with the growth of TGO [51–53]. With the growth of the TGO, trace amounts of Cr^{3+} ions are introduced into α-Al$_2$O$_3$. When the position of the Al^{3+} ion at the center of the octahedral α-Al$_2$O$_3$ cell is replaced by a Cr^{3+} ion [54], two emission peaks near 693 nm are generated from the energy transition of ^{2}E-^{4}A$_{2g}$ under laser irradiation. These two peaks are defined as R_1 and R_2 at wavenumbers of approximately 14,403 cm^{-1} (694.30 nm) and 14,433 cm^{-1} (692.86 nm), respectively. Although the α-Al$_2$O$_3$ in the TGO is usually polycrystalline, the line type of the Cr^{3+} fluorescence characteristic peak was more similar to the Lorentzian line type because of the single mechanism of fluorescence generation, which is confirmed by the fitting phenomena shown in Figure 4 and the results listed in Table 1.

In addition, multipeak fitting and the barycenter method were also applied for the identification of peak shifts in some published works [55]. The linear superposition of multiple single-peak functions makes the fitted curve closely fit the measured data, achieving an extremely high GOF of more than 0.999. However, the fitting parameters of the multipeak functions have neither physical significance nor enough stability. Under the same parameter settings, the fitting results at different times are different. They are sensitive to the number of preselected peaks, as shown in Figure 11. The barycenter method takes the barycentric wavenumber of a specific spectral band as its peak position, which is suitable for spectral bands with poor symmetry, such as Eu^{3+} fluorescence peaks [32]. Neither of the above methods are suitable for the data processing of Cr^{3+} fluorescence spectra.

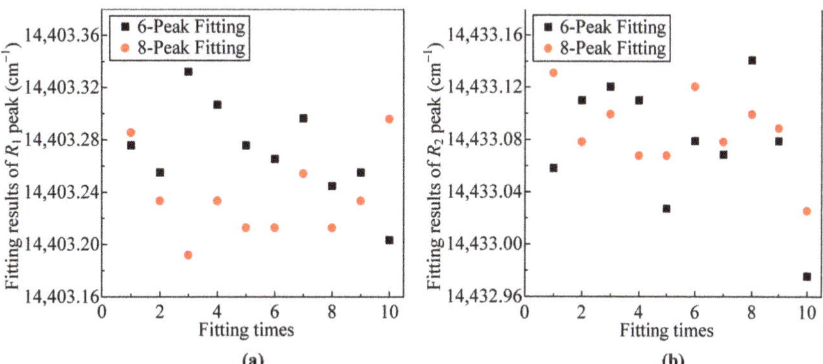

Figure 11. Fitting results of identifying peak positions by the multipeak fitting method: (**a**) results of R_1; (**b**) results of R_2.

The bullet point list of our key findings above is as follows:

- The fitting results show that the Psd-Voigt function is not sensitive enough to small stress when fitting the Cr^{3+} photoluminescence spectrum, even though its GOF is relatively high. Both Gaussian and Lorentzian functions can accurately identify the peak shifts corresponding to different stresses.
- The unilateral pursuit of Psd-Voigt function fitting process is to achieve a GOF that is as high as possible by adjusting the Lorentzian–Gaussian weight ratio, which has no physical meaning. Due to the loss of a unified standard for mutual comparison, the Psd-Voigt function is insensitive and inaccurate for stress analysis.
- The applicability of any fitting function to spectral analysis is determined by the generation mechanism of the spectrum. The line type of the Cr^{3+} fluorescence characteristic peak was more similar to the Lorentzian line type because of the single mechanism of fluorescence generation, which is confirmed by the fitting phenomena and the results.
- Comparatively speaking, Lorentzian fitting is proven to be the most applicable method.

5. Conclusions

Methods used to process Cr^{3+} photoluminescence spectroscopic data were studied to analyze the application of stress on the thermally grown oxide layers of TBCs. Numerical experiments and calibration tests were carried out to obtain numerous spectra corresponding to different uniaxial loadings. The results showed that the Psd-Voigt fitting obtained the highest goodness-of-fit, but it had low sensitivity for small stress changes, as well as low stability and confidence of peak shifts, which indicated that it was far from suitable for analyzing stress by Cr^{3+} photoluminescence piezospectroscopy, even though it is very widely utilized in published works. Lorentzian fitting was proven to be the most applicable method, as shown by comparing its fitting results with those of other functions and then discussing the generation mechanism of Cr^{3+} photoluminescence piezospectroscopy.

Author Contributions: Conceptualization, W.Q.; methodology, Y.Z. and N.L.; formal analysis, Y.Z. and N.L.; investigation, Y.Z. and N.L.; resources, W.Q.; data curation, Y.Z. and N.L.; writing—original draft preparation, N.L. and Y.Z.; writing–reviewing and editing, W.Q., Y.Z., and N.L.; visualization, W.Q., Y.Z., and N.L.; supervision, W.Q. and Y.Z.; project administration, W.Q.; funding acquisition, W.Q. All authors have read and agreed to the published version of the manuscript.

Funding: This research was funded by the National Natural Science Foundation of China, Grant Nos. 12021002, 11827802, 11772223, 11772227 and 11890682.

Institutional Review Board Statement: Not applicable.

Informed Consent Statement: Not applicable.

Data Availability Statement: The data are not publicly available due to that the data also forms part of an ongoing study.

Conflicts of Interest: The authors declare no conflict of interest.

References

1. Clarke, D.R.; Oechsner, M.; Padture, N.P. Thermal-barrier coatings for more efficient gas-turbine engines. *MRS Bull.* **2012**, *37*, 891–899. [CrossRef]
2. Hardwicke, C.U.; Lau, Y.C. Advances in thermal spray coatings for gas turbines and energy generation: A review. *J. Therm. Spray Technol.* **2013**, *22*, 564–576. [CrossRef]
3. Clarke, D.R.; Phillpot, S.R. Thermal barrier coating materials. *Mater. Today* **2005**, *8*, 22–29. [CrossRef]
4. Jiang, P.; Fan, X.; Sun, Y. Bending-driven failure mechanism and modelling of double-ceramic-layer thermal barrier coating system. *Int. J. Solids Struct.* **2017**, *130*, 11–20. [CrossRef]
5. Miller, R.A. Oxidation-based model for thermal barrier coating life. *J. Am. Ceram. Soc.* **1984**, *67*, 517–521. [CrossRef]
6. Widjaja, S.; Limarga, A.M.; Yip, T.H. Modeling of residual stresses in a plasma-sprayed zirconia/alumina functionally graded-thermal barrier coating. *Thin Solid Films* **2003**, *434*, 216–227. [CrossRef]
7. Padture, N.P. Thermal barrier coatings for gas-turbine engine applications. *Science* **2002**, *296*, 280–284. [CrossRef] [PubMed]
8. Wellman, R.G.; Nicholls, J.R. Erosion, corrosion and erosion–corrosion of EB PVD thermal barrier coatings. *Tribol. Int.* **2008**, *41*, 657–662. [CrossRef]
9. Evans, H.E. Oxidation failure of TBC systems: An assessment of mechanisms. *Surf. Coat. Technol.* **2011**, *206*, 1512–1521. [CrossRef]
10. Bengtsson, P.; Persson, C. Modelled and measured residual stresses in plasma sprayed thermal barrier coatings. *Surf. Coat. Technol.* **1997**, *92*, 78–86. [CrossRef]
11. Jiang, P.; Fan, X.; Sun, Y. Thermal-cycle dependent residual stress within the crack-susceptible zone in thermal barrier coating system. *J. Am. Ceram. Soc.* **2018**, *101*, 4256–4261. [CrossRef]
12. Guo, H.; Kuroda, S.; Murakami, H. Microstructures and properties of plasma-sprayed segmented thermal barrier coatings. *J. Am. Ceram. Soc.* **2006**, *89*, 1432–1439. [CrossRef]
13. Civera, M.; Fragonara, L.Z.; Surace, C. An experimental study of the feasibility of phase-based video magnification for damage detection and localisation in operational deflection shapes. *Strain* **2020**, *56*, e12336. [CrossRef]
14. Lima, C.R.C.; Dosta, S.; Guilemany, J.M. The application of photoluminescence piezospectroscopy for residual stresses measurement in thermally sprayed TBCs. *Surf. Coat. Technol.* **2017**, *318*, 147–156. [CrossRef]
15. Sohn, Y.H.; Schlichting, K.; Vaidyanathan, K. Nondestructive evaluation of residual stress for thermal barrier coated turbine blades by Cr^{3+} photoluminescence piezospectroscopy. *Metall. Mater. Trans. A* **2000**, *31*, 2388–2391. [CrossRef]
16. Selcuk, A.; Atkinson, A. The evolution of residual stress in the thermally grown oxide on Pt diffusion bond coats in TBCs. *Acta Mater.* **2003**, *51*, 535–549. [CrossRef]

17. Chen, C.; Guo, H.; Gong, S. Sintering of electron beam physical vapor deposited thermal barrier coatings under flame shock. *Ceram. Int.* **2013**, *39*, 5093–5102. [CrossRef]

18. Ma, Q.; Clarke, D.R. Stress measurement in single-crystal and polycrystalline ceramics using their optical fluorescence. *J. Am. Ceram. Soc.* **1993**, *76*, 1433–1440. [CrossRef]

19. Christensen, R.J.; Lipkin, D.M.; Clarke, D.R. Nondestructive evaluation of the oxidation stresses through thermal barrier coatings using Cr^{3+} piezospectroscopy. *Appl. Phys. Lett.* **1996**, *69*, 3754–3756. [CrossRef]

20. Nychka, J.A.; Clarke, D.R. Damage quantification in TBCs by photo-stimulated luminescence spectroscopy. *Surf. Coat. Technol.* **2001**, *146*, 110–116. [CrossRef]

21. Schlichting, K.W.; Vaidyanathan, K.; Sohn, Y.H.; Jordan, E.H.; Gell, M.; Padture, N.P. Application of Cr^{3+} photoluminescence piezo-spectroscopy to plasma-sprayed thermal barrier coatings for residual stress measurement. *Mater. Sci. Eng. A* **2000**, *291*, 68–77. [CrossRef]

22. Shen, Z.; Zhong, B.; He, L.; Xu, Z.; Mu, R.; Huang, G. LaZrCeO/YSZ TBCs by EB-PVD: Microstructural evolution, residual stress and degradation behavior. *Mater. Chem. Front.* **2019**, *3*, 892–900. [CrossRef]

23. Jiang, P.; Yang, L.; Sun, Y.; Li, D.; Wang, T. Nondestructive measurements of residual stress in air plasma-sprayed thermal barrier coatings. *J. Am. Ceram. Soc.* **2020**. [CrossRef]

24. Clarke, D.R.; Christensen, R.J.; Tolpygo, V. The evolution of oxidation stresses in zirconia thermal barrier coated superalloy leading to spalling failure. *Surf. Coat. Technol.* **1997**, *94–95*, 89–93. [CrossRef]

25. Selcuk, A.; Atkinson, A. Analysis of the Cr^{3+} luminescence spectra from thermally grown oxide in thermal barrier coatings. *Mater. Eng.* **2002**, *335*, 147–156. [CrossRef]

26. Manero, A.; Selimov, A.; Wischek, J.; Fouliard, Q.; Meid, C. Piezospectroscopic evaluation and damage identification for thermal barrier coatings subjected to simulated engine environments. *Surf. Coat. Technol.* **2017**, *323*, 30–38. [CrossRef]

27. Zhu, W.; Zhang, Z.B.; Yang, L.; Zhou, Y.C.; Wei, Y.G. Spallation of thermal barrier coatings with real thermally grown oxide morphology under thermal stress. *Mater. Des.* **2018**, *146*, 180–193. [CrossRef]

28. Wei, Z.Y.; Cai, H.N.; Li, C.J. Comprehensive dynamic failure mechanism of thermal barrier coatings based on a novel crack propagation and TGO growth coupling model. *Ceram. Int.* **2018**, *44*, 22556–22566. [CrossRef]

29. Wei, Z.Y.; Cai, H.N.; Meng, G.H.; Tahir, A.; Zhang, W.W. An innovative model coupling TGO growth and crack propagation for the failure assessment of lamellar structured thermal barrier coatings. *Ceram. Int.* **2019**, *46*, 1532–1544. [CrossRef]

30. Fabisiak, K.; Los, S.; Paprocki, K.; Szybowicz, M.; Dychalska, A. Orientation dependence of cathodoluminescence and photoluminescence spectroscopy of defects in chemical-vapor-deposited diamond microcrystal. *Materials* **2020**, *13*, 5446. [CrossRef]

31. Sonntag, J.; Li, J.; Plaud, A.; Loiseau, A.; Barjon, J.; Edgar, J.H.; Stampfer, C. Excellent electronic transport in heterostructures of graphene and monoisotopic boron-nitride grown at atmospheric pressure. *2D Mater.* **2020**, *7*, 031009. [CrossRef]

32. Zhao, S.; Ren, Z.; Zhao, Y.; Xu, J.; Zou, B.; Hui, Y.; Zhu, L.; Zhou, X.; Cao, X. The application of Eu^{3+} photoluminescence piezo-spectroscopy in the $LaMgAl_{11}O_{19}$/8YSZ:Eu double-ceramic-layer coating system. *J. Eur. Ceram. Soc.* **2015**, *35*, 249–257. [CrossRef]

33. Li, Q.; Qiu, W.; Tan, H.; Guo, J.; Kang, Y. Micro-Raman spectroscopy stress measurement method for porous silicon film. *Opt. Lasers Eng.* **2010**, *48*, 1119–1125. [CrossRef]

34. Xu, C.; Xue, T.; Qiu, W.; Kang, Y. Size effect of the interfacial mechanical behavior of graphene on a stretchable substrate. *ACS Appl. Mater. Interfaces* **2016**, *8*, 27099–27106. [CrossRef] [PubMed]

35. Lei, Z.K.; Kang, Y.L.; Cen, H.; Hu, M.; Qiu, Y. Residual stress on surface and cross-section of porous silicon studied by micro-Raman spectroscopy. *Chin. Phys. Lett.* **2005**, *22*, 984–986. [CrossRef]

36. Qiu, W.; Cheng, C.L.; Liang, R.R.; Zhao, C.W.; Lei, Z.K.; Zhao, Y.C.; Ma, L.L.; Xu, J.; Fang, H.J.; Kang, Y.L. Measurement of residual stress in a multi-layer semiconductor heterostructure by micro-Raman spectroscopy. *Acta Mech. Sin.* **2016**, *32*, 805–812. [CrossRef]

37. Dou, W.; Xu, C.; Guo, J.; Du, H.; Qiu, W.; Xue, T.; Kang, Y.; Zhang, Q. Interfacial mechanical properties of double-layer graphene with consideration of the effect of stacking mode. *ACS Appl. Mater. Interfaces* **2018**, *10*, 44941–44949. [CrossRef]

38. Qiu, W.; Ma, L.; Li, Q.; Xing, H.; Cheng, C.; Huang, G. A general metrology of stress on crystalline silicon with random crystal plane by using micro-Raman spectroscopy. *Acta Mech. Sin.* **2018**, *34*, 1095–1107. [CrossRef]

39. Syassen, K. Ruby under pressure. *High Pressure Res.* **2008**, *28*, 75–126. [CrossRef]

40. Moss, S.C.; Newnham, R. The chromium position in ruby. *Z. Kris.* **1964**, *120*, 359–363. [CrossRef]

41. Grabner, L. Spectroscopic technique for the measurement of residual stress in sintered Al_2O_3. *J. Appl. Phys.* **1978**, *49*, 580–583. [CrossRef]

42. Forman, R.A.; Piermarini, G.J.; Barnett, J.D.; Block, S. Pressure measurement made by the utilization of ruby sharp-line luminescence. *Science* **1972**, *176*, 284–285. [CrossRef] [PubMed]

43. Piermarini, G.J.; Block, S.; Barnett, J.; Forman, R. Calibration of the pressure dependence of the R_1 ruby fluorescence line to 195 kbar. *J. Appl. Phys.* **1975**, *46*, 2774–2780. [CrossRef]

44. Eggert, J.H.; Goettel, K.A.; Silvera, I.F. Ruby at high pressure. I. Optical line shifts to 156 GPa. *Phys. Rev. B* **1989**, *40*, 5724–5732. [CrossRef] [PubMed]

45. McClure, D.S. Optical spectra of transition-metal ions in corundum. *J. Chem. Phys.* **1962**, *36*, 2757–2779. [CrossRef]

46. Sugano, S.; Tsujikawa, I. Absorption spectra of Cr^{3+} in Al_2O_3 Part B. experimental studies of the Zeeman effect and other properties of the line spectra. *J. Phys. Soc. Jpn.* **1958**, *13*, 899–910. [CrossRef]

47. Fukuchi, T.; Eto, S.; Okada, M.; Fujii, T. Evaluation of applicability of compact excitation light source to detection of thermally grown oxide layer in thermal barrier coating for gas turbines. *Electr. Eng. Jpn.* **2015**, *190*, 1–8. [CrossRef]
48. Wang, X.; Lee, G.; Atkinson, A. Investigation of TBCs on turbine blades by photoluminescence piezospectroscopy. *Acta Mater.* **2009**, *57*, 182–195. [CrossRef]
49. Dassios, K.G.; Galiotis, C. Fluorescence studies of polycrystalline Al_2O_3 composite constituents: Piezo-spectroscopic calibration and applications. *Appl. Phys. A Mater.* **2004**, *79*, 647–659. [CrossRef]
50. Huang, M.; Yan, H.; Chen, C.; Song, D.; Heinz, T.F.; Hone, J. Phonon softening and crystallographic orientation of strained graphene studied by Raman spectroscopy. *Proc. Natl. Acad. Sci. USA* **2009**, *106*, 7304–7308. [CrossRef] [PubMed]
51. Sun, Y.; Li, J.; Zhang, W.; Wang, T.J. Local stress evolution in thermal barrier coating system during isothermal growth of irregular oxide layer. *Surf. Coat. Technol.* **2013**, *216*, 237–250. [CrossRef]
52. Sun, Y.; Zhang, W.; Li, J.; Wang, T. Local stress around cap-like portions of anisotropically and nonuniformly grown oxide layer in thermal barrier coating system. *J. Mater. Sci.* **2013**, *48*, 5962–5982. [CrossRef]
53. Sridharan, S.; Xie, L.D.; Jordan, E.H.; Gell, M. Stress variation with thermal cycling in the thermally grown oxide of an EB-PVD thermal barrier coating. *Surf. Coat. Technol.* **2004**, *179*, 286–296. [CrossRef]
54. Kenyon, P.; Andrews, L.; McCollum, B.; Lempicki, A. Tunable infrared solid-state laser materials based on Cr^{3+} in low ligand fields. *IEEE J. Quantum Electron.* **1982**, *18*, 1189–1197. [CrossRef]
55. Zhang, W.; Li, Z.; Baxter, G.W.; Collins, S.F. Stress- and temperature-dependent wideband fluorescence of a phosphor composite for sensing applications. *Exp. Mech.* **2017**, *57*, 57–63. [CrossRef]

Article

Measurement of Stress Optic Coefficient for Thermal Barrier Coating Based on Terahertz Time-Domain Spectrum

Zong Wang [1,†], Yanheng Zhang [1,†], Ning Lu [1], Zhiyong Wang [1,2,*] and Wei Qiu [1,2,*]

[1] Department of Mechanics, Tianjin University, Tianjin 300354, China; wangz2020@tju.edu.cn (Z.W.); tjuzhangyh@tju.edu.cn (Y.Z.); sugariss@tju.edu.cn (N.L.)
[2] Tianjin Key Laboratory of Modern Engineering Mechanics, Tianjin University, Tianjin 300354, China
* Correspondence: zywang@tju.edu.cn (Z.W.); qiuwei@tju.edu.cn (W.Q.)
† Co-first author.

Abstract: The residual stress introduced inside the thermal barrier coating (TBC) top coating during manufacturing and service processes is one of the main causes of thermal barrier failure. Therefore, a nondestructive and accurate measurement of the residual stress in top coating is essential for the evaluation of TBC life. The terahertz time-domain spectroscopy (THz-TDS) technique, which is based on the calibration or measurement of the stress optical coefficients of the measured materials, is applicable to the measuring of internal stress of nonmetal materials. In this work, to characterize the internal stress in TBC, the stress optic coefficient of the TBC top coating was measured by reflection-type THz-TDS. First, the mechanics model for the internal stress measurement in a TBC top coating was derived based on the photoelastic theory. Then, the THz time-domain spectra of TBC specimens under different loadings were measured in situ by a reflection-type THz-TDS system. Finally, the unimodal fitting, multimodal fitting and barycenter methods were used to carry out the data processing of the THz time-domain spectral-characteristic peaks. By comparing the processed results, the results using the barycenter method were regarded as the calibrated stress optical coefficient of the TBC due to the method's sufficient accuracy and stability.

Keywords: thermal barrier coating; terahertz time-domain spectroscopy (THz-TDS); stress optic coefficient; barycenter method

Citation: Wang, Z.; Zhang, Y.; Lu, N.; Wang, Z.; Qiu, W. Measurement of Stress Optic Coefficient for Thermal Barrier Coating Based on Terahertz Time-Domain Spectrum. *Coatings* **2021**, *11*, 1265. https://doi.org/10.3390/coatings11101265

Academic Editor: Narottam P. Bansal

Received: 13 September 2021
Accepted: 13 October 2021
Published: 18 October 2021

Publisher's Note: MDPI stays neutral with regard to jurisdictional claims in published maps and institutional affiliations.

1. Introduction

The gas turbine is one of the core pieces of power equipment closely related to the development of the energy and marine industries. The gas temperature in the front of the turbine blade is one of the indicators of the technical level of the gas turbine, which can reach more than 1600 °C. A thermal barrier coating (TBC) is one of the effective means to raise the gas temperature of a gas turbine due to its low cost and excellent comprehensive performance [1–4]. A TBC is a thermal protective multilayer coating structure formed by spraying materials with low thermal conductivity and high thermal stability onto metal substrates, and mainly includes top coating (TC), bonding coating (BC) and metal substrate. In addition, thermal growth oxide (TGO) may be generated when TBC works at high temperatures. A TBC has a typical multilayer heterostructure; hence, process stress and service stress are inevitably introduced into the coatings during the manufacturing process and by the service environment. For instance, internal stress is generated inside the top coating during the entire service life, owing to the microstructural evolution and thermal mismatch [5–7]. Interfacial stress is introduced with the growth of TGO. These internal and interfacial stresses may cause cracks and spalls along the interfaces and/or throughout the layers of the TBC, leading to the structural failure of the whole TBC system, and thus seriously threatening the service safety of the gas turbine [8–12]. Therefore, for the service safety of the gas turbine, it is important to characterize the residual stress in TBC using a nondestructive and accurate method.

For the nondestructive measurement of stress in the TBC top coating, many experimental techniques have been developed, such as X-ray diffraction (XRD), micro-Raman spectroscopy (μRS) and photoluminescence spectroscopy (PL). The XRD technique [13–16] characterizes the stress using the lattice deformation it causes, which has a high measurement accuracy. The μRS technique [17–19] measures stress with a high resolution through the detection of the frequency shift of the characteristic peak of the Raman spectrum line caused by crystal lattice deformation. However, XRD and μRS techniques are limited to surface stress measurements and by low measurement efficiency. Based on the photoluminescence piezo-spectroscopy effect, the PL technique [20–22] measures the stress at a certain depth of the TBC top coating instead of that in the whole TC layer of TBC. In addition, the resolution and confidence of the existing PL technique to characterize the internal stress of the TBC top coating are approximately several hundred MPa. Hence, existing nondestructive measurement technologies cannot meet the experimental requirements of high precision, high efficiency and full-field stress measurements.

Terahertz time-domain spectroscopy (THz-TDS) has recently emerged as a novel nondestructive testing method with the advantages of convenience, noncontact, and high precision. Most opaque nonmetallic materials, including the top coating materials of the TBC [23–25], have birefringence properties when penetrated by terahertz (THz) waves, which make it possible to quantify the internal stress based on the principle of photoelasticity [26–29]. Based on a reflection-type THz-TDS system, Chen et al. found that THz waves could penetrate the TBC top coating and reflect back at both the air/TC interface and the TC/BC interface [30]. The transmission-type THz-TDS system was used to measure the transmittance of TBC by Watanabe et al., who found that the coating had a good transmittance of 20% to 80% in the frequency range of less than 0.5 THz [31]. This result indicated that THz-TDS could be used for the measurement of the TBC structure. Based on a transmission-type THz-TDS system, the stress optic coefficient of sintered yttria partially stabilized zirconia (YTZP) ceramic bulk in a tensional state was measured by Schemmel et al. [32]. In this experiment, the phenomenon of stress-induced birefringence in a ceramic under tension was observed, which proved the feasibility of the quantitative measurement of internal stress in a ceramic coating using THz-TDS. Schemmel et al. also extended the THz-TDS study on the TBC to a self-built reflection-type THz-TDS system, which also found the stress-induced birefringence phenomenon in TBC under tension-loading [33].

In fact, a transmission-type THz-TDS system is hardly applicable for the analysis of stress in an actual TBC structure because the metal substrate of TBC is opaque to the THz 'light'. Hence, a reflection-type THz-TDS system is essential for the stress measurement of an actual TBC.

The kernel problem is the stress optical coefficient of the measured material. The TC layer in TBC is generally a specific ceramic material, such as rare metal-doped zirconia, which is porous, brittle and prepared by means of atmospheric plasma spraying (APS) or electron beam-physical vapor deposition (EB-PVD). The mechanical properties of the TC layer, including its stress optical coefficient, vary greatly due to different stress states (tension or compression), preparation methods and production batches. In particular, in most cases, the stress state of the TBC top coating is compressive stress. The stress optical coefficient under tension in most of existing works was not helpful enough in the actual analysis of the TC layer under compressive stress.

In the present work, the stress optical coefficient of sintered 8 wt.% yttrium oxide partially stabilized zirconia under compression was measured based on a reflection-type THz-TDS system and a uniaxial loading device. Based on the unimodal fitting, the multimodal fitting and the barycenter methods, the stress optical coefficient of the TBC top coating was calibrated by fitting the slope between the refractive index and the stress of the TBC top coating, which is the significant parameter in THz stress analysis. The measurement of the stress optical coefficient provides the basis for the internal stress analysis of the TBC top coating of a gas turbine using THz-TDS.

2. Materials and Experiments

The specimens used in this work were prepared as follows. Sintered 8 wt.% yttrium oxide partially stabilized zirconia (8YSZ) powder was put into a mold and pressed into the ceramic embryo body. The 8YSZ powder was supplied by Tianjin Detianzhu Amorphous Nano Technology Ltd. (Tianjin, China). Then, the embryo body was sintered at 1500 °C to form a dense 8YSZ ceramic block. After that, the TBC specimens with high surface flatness were obtained by high-precision wire cutting and fine grinding. Eventually, one side of each specimen was coated with a 1 μm aluminum film using the radio frequency magnetron sputtering system (YCL-50A, Yujie Vacuum Equipment Ltd., Shenyang, China). The uniform thickness of the aluminum film was ensured by rotating the TBC specimen. The power of the equipment was 60 W, the working pressure was 2 Pa, and the deposition time was 2 h. The final dimensions of each specimen are 15 mm × 10 mm × 1 mm. In addition, the specimen has the same composition and preparation process as the actual TBC top coating.

The experimental installation, as shown in Figure 1, consisted of a reflection-type THz-TDS system and a uniaxial loading machine. In detail, the THz-TDS system was an all-fiberized terahertz time-domain spectrometer (TP15K-F, produced by Tianjin Tiertime Ltd., Tianjin, China). Two all-fiberized THz antennas were chosen to generate and detect the THz waves. The femtosecond laser had a 100-fs pulse width, a 1550 nm wavelength, and a 100 MHz repetition rate. The frequency range of available THz pulses in the system was 0.2 to 3.5 THz. The THz wave was focused onto the specimen using a convex lens, and the spot diameter on the specimen was approximately 5 mm. The incidence angle of the THz wave was 45°. In addition, a self-built uniaxial loading machine was used for the experiments.

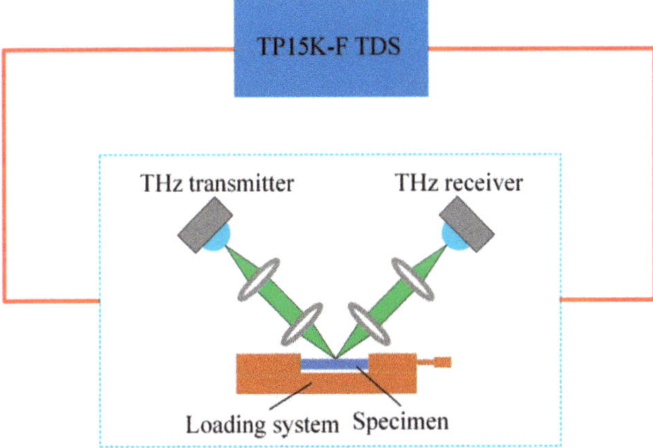

Figure 1. Diagram of the experimental installation for the measurement of the stress optical coefficient.

Before the test, the TBC specimen was fixed on the loading machine using a compressive preload of 30 N and adjusted to ensure that the axis of the specimen along the length direction was perpendicular to the contact surface of the loading block. Then, the specimen was compressively loaded step-by-step to 800 N with a step length of 50 N. The reflective THz time-domain spectra were obtained by averaging 100 in situ collections under each loading step. The THz time-domain spectra of TBC specimens under different loadings collected in this work are shown in Figure 2, where S denotes surface reflection, R1 denotes the first interface reflection, and R2 denotes the second interface reflection.

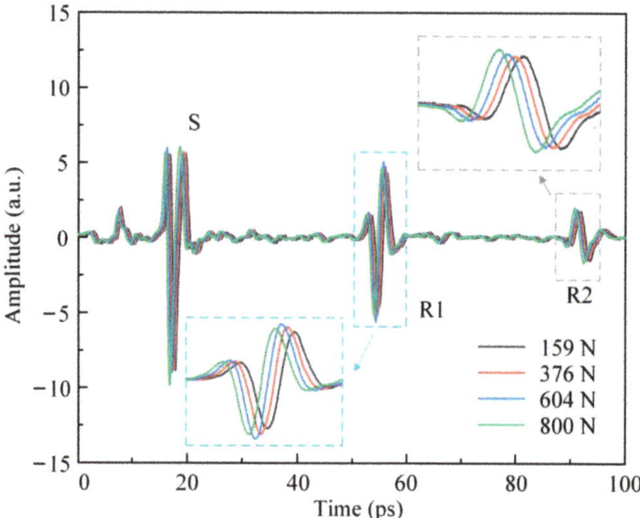

Figure 2. Diagram of the reflective terahertz time-domain spectroscopy.

3. Models and Methods

3.1. Model of Stress Measurement

The THz wave of the ceramic material in a TBC has a sensitive stress-induced birefringence effect [32]. Assuming that the stress in the TBC top coating conforms to the in-plane stress state, the birefringence phenomenon will occur when the beam penetrates vertically into the material. The beam transmitting into the TC layer is decomposed into two plane-polarized beams perpendicular to each other along the principal stress direction, and the two beams of polarized light propagate at different speeds in the ceramic material. The relationship between the principal stress of the TC layer, σ_1, σ_2, and the refractive index in the direction of the principal stress, n_1, n_2, can be expressed as follows:

$$\begin{cases} n_1 - n = c_1\sigma_1 + c_2\sigma_2 \\ n_2 - n = c_1\sigma_2 + c_2\sigma_1 \end{cases} \tag{1}$$

where n denotes the refractive index of the unloaded material, n_1 and n_2 denote the absolute refractive index of the ceramic material in the directions of σ_1 and σ_2, and c_1 and c_2 denote the stress optical coefficients in the directions of σ_1 and σ_2. By subtracting the two expressions in Equation (1), it can be stated that:

$$n_1 - n_2 = C(\sigma_1 - \sigma_2) \tag{2}$$

where C denotes the stress optical coefficient of the material and $C = c_1 - c_2$. If a uniaxial load is applied such that the transverse stress σ_2 is approximately zero, Equation (1) can be expressed as:

$$\Delta n_1 = c_1 \times \Delta\sigma_1 \tag{3}$$

The stress optical coefficient c_1 of the material can be obtained by calculating the refractive index change of the specimen under different loadings, which can be used to measure the internal stress of the TBC top coating. Hence, the measurement of the stress optical coefficient of the TBC material is the basis of the stress analysis in the TBC top coating.

3.2. Model of Refractive Index Measurement

THz waves can penetrate the ceramic, but not the metal, and are reflected back at both the air/TC interface and the TC/BC interface (viz. the 8YSZ/Al interface). The propagation path of the THz wave inside the ceramic TC is schematically shown in Figure 3, where Δt denotes the time delay of THz wave, c denotes the speed of light in a vacuum and d denotes the thickness of the specimen. In a reflection-type THz-TDS system, the relationship between the specimen refractive index n_{TC} and time delay Δt can be expressed as [34].

$$n_{TC} = \frac{c \times \Delta t}{2d} \tag{4}$$

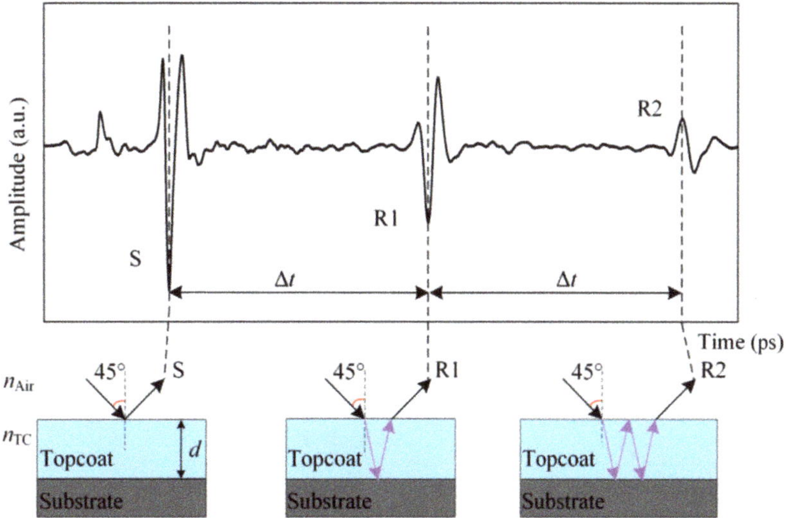

Figure 3. Schematic diagram of the reflective terahertz propagation path inside a specimen.

3.3. Data Processing Methods

In this work, the positions of characteristic peaks were extracted from the THz time-domain spectra using three data processing methods: a unimodal fitting method, a multimodal fitting method and a barycenter method. The Lorentz function was used for unimodal fitting because the characteristic peaks of the THz spectra were narrow and sharp enough. The Lorentz function is expressed as follows:

$$y = y_0 + \frac{2A}{\pi} \times \frac{w}{4(x - x_c)^2 + w^2} \tag{5}$$

where y_0 denotes the reference value of the characteristic peak, A denotes the peak area, w denotes the half-height width, and x_c denotes the peak position.

The Lorentz multimodal fitting method extracted the positions of the characteristic peaks through the linear superposition of multiple Lorentz functions. Fitting curves with a high goodness of fit (GOF) to the original data were obtained by using multimodal fitting. The spectral GOF obtained using multimodal fitting was relatively high (up to 0.9999); that is, the fitting curves almost perfectly matched the experimental curves.

In addition, the barycenter method was also applied to process the measured spectra in the present work. The peak position corresponding to the half-peak-area of the characteristic peak (the barycentric peak) was used in the barycenter method to describe the spectral deviation. The THz time-domain spectrum is given in Figure 4a, which shows that the whole characteristic peak can be divided into two regions, the left region and the right region, by a vertical line. When the integral of the left region was equal to that of the right region, as shown in Figure 4b, the position of the vertical line dividing the two regions was regarded as the barycentric peak.

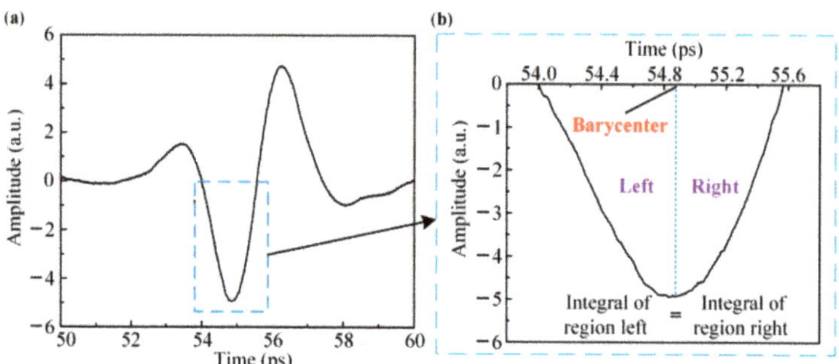

Figure 4. Schematic diagram of the barycenter method. (**a**) the selected THz time-domain spectrum; (**b**) the position of the barycentric peak.

4. Results and Discussions

In this work, 15 groups of time-domain spectra under different stress steps were obtained using a reflection-type THz-TDS system and processed by using unimodal fitting, multimodal fitting and barycenter methods, respectively. The refractive indices under different uniaxial loadings were obtained, as shown in Figure 5. Based on the processed results, the stress optical coefficients c_1 were obtained and are listed in Table 1.

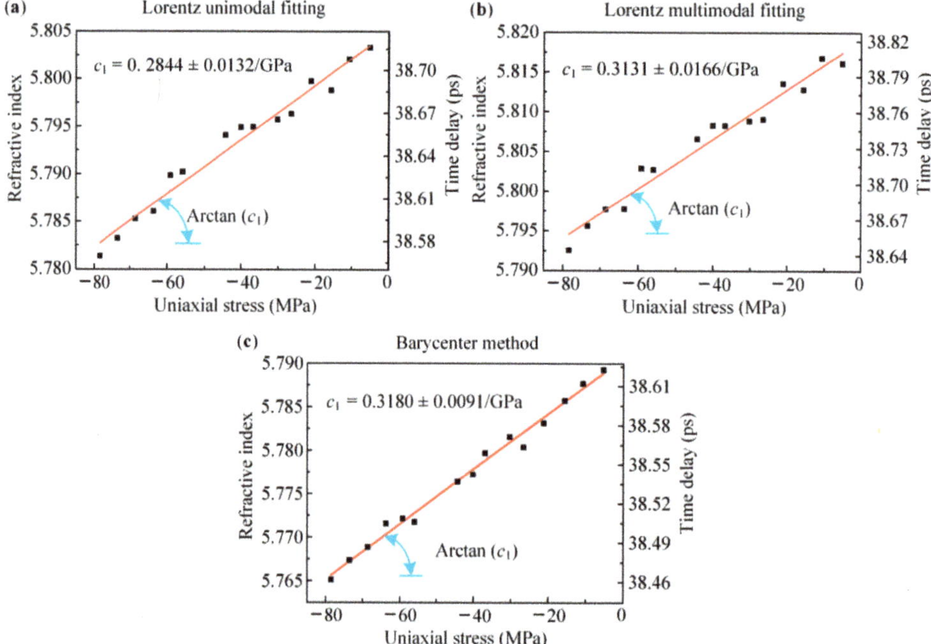

Figure 5. The relation between the refractive index and stress under using different processing methods: (**a**) unimodal fitting method; (**b**) multimodal fitting method; (**c**) barycenter method.

Table 1. Processing results for measurement experiments.

Method	Stress Optical Coefficient c_1 (GPa)	Linear Goodness of Fit
Unimodal fitting method	0.2844 ± 0.0132	0.9728
Multimodal fitting method	0.3131 ± 0.0166	0.9648
Barycenter method	0.3180 ± 0.0091	0.9894

In Figure 5, it can be seen that there is always a linear relationship between the refractive index of the TBC specimen and the internal stress within the online elastic range. This indicates the existence of stress-induced birefringence in the TBC specimen, which verifies the feasibility of measuring the internal stress of the TBC top coating using the THz-TDS technique. In addition, the stress optical coefficient can be obtained by using the above three fitting methods.

By comparing the results of these three methods, the stress optical coefficient obtained from the multimodal fitting was found to be similar to that of the barycenter method, but that of the unimodal fitting was significantly lower than the other two, since the Lorentz curve is left–right symmetric and the characteristic peak in the THz time-domain spectrum is scarcely symmetric. As shown in Figure 6, the unimodal fitting curve was different from the original curve in shape, leading to a low peak GOF (0.9922). The width of the left side of the characteristic peak was wider than that of the right side, which lead the peak position of the fitting curve to deviate to the left during unimodal fitting. In Figure 6, the peak position of the original curve was approximately 54.528 ps and that of the unimodal fitting curve was 54.4945 ps. The average time-shift of the peak position of the characteristic peak under each loading step was 0.0815 ps, and the deviation calculated from the unimodal fitting was 0.0335 ps. The calculation error of the peak position (the ratio of the deviation of the peak position to the average time-shift) was 41.1%. In addition, the unimodal fitting results of the THz spectrum of TBC specimens under different loadings were processed by the above method and the average calculation error of the unimodal fitting method was 28.2%. Hence, the Lorentz unimodal fitting method was not suitable due to the poor symmetry of the data detected by THz-TDS.

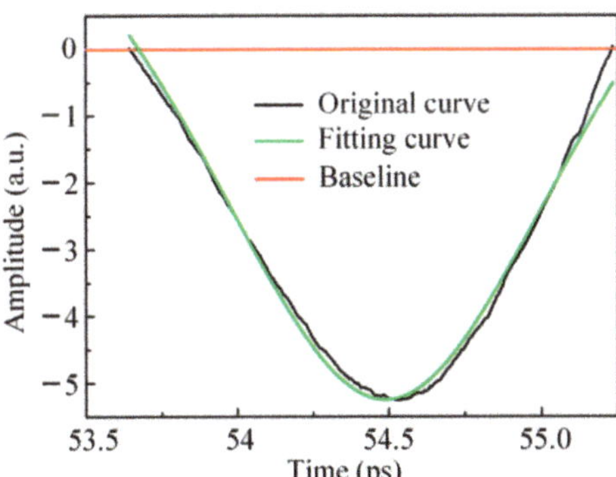

Figure 6. Result of unimodal fitting for terahertz time-domain spectroscopy.

For the multimodal fitting method, the number of sub-peaks and their positions had a great influence on the fitting results. In this work, multimodal fitting was performed for the THz time-domain spectrum based on different numbers of sub-peaks. As shown in Figure 7a, when the number of sub-peaks was five, the fitting curve was basically consistent with the original curve, but not in detail. The peak GOF was 0.9997, the deviation of the peak position was 0.0020 ps, and the calculation error was 2.4%. The peak position accuracy of the multimodal fitting method was higher than that of the unimodal fitting method. In Figure 7b, when the number of sub-peaks was 10, the fitting curve coincided with the height of the original curve, and the peak GOF was 0.9999. The deviation of the peak position was 0.0003 ps, and the calculation error was 0.4%. Therefore, with the increase in the number of sub-peaks, the fitting peak increasingly overlapped with the characteristic peak, and the accuracy of fitting the peak position was continuously improved. However, as shown in Table 1, the linear GOF of the multimodal fitting method was 0.9661, which was the lowest among the three methods. This is because the multimodal fitting method unilaterally pursues high-peak GOF and ignores partial spectral information, resulting in a reduction in calibration accuracy. In addition, it is difficult to determine the number and position of the sub-peaks that constitute the characteristic peaks, leading to a low success rate of multimodal fitting. Therefore, the multimodal fitting method is greatly affected by human factors, and the fitting results are less stable.

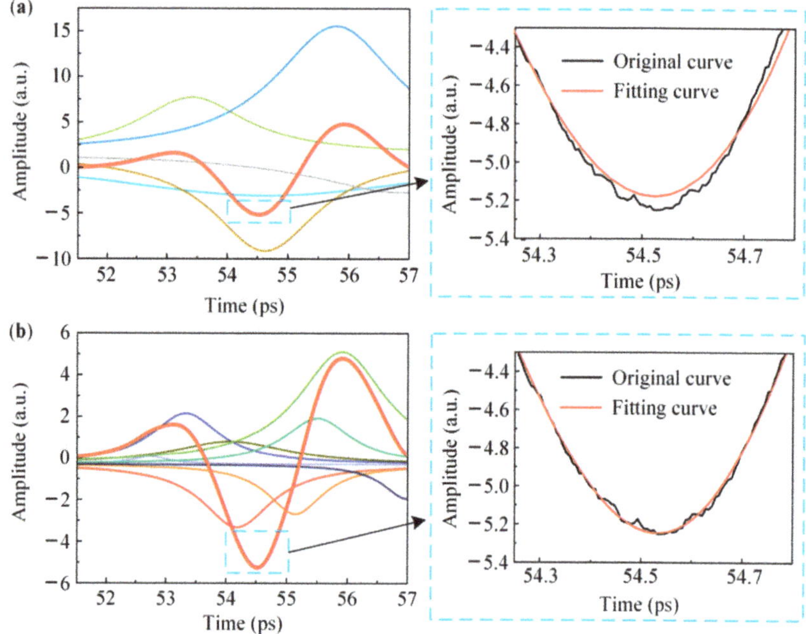

Figure 7. Result of multimodal fitting for terahertz time-domain spectroscopy with the comparison with original curve: (**a**) the number of sub-peaks is 5; (**b**) the number of sub-peaks is 10.

Spectral analysis using a Psd-Voigt function or multimodal fitting always obtains a relatively optimal GOF. However, the unilateral pursuit of such a fitting process is to achieve a GOF that is as high as possible by adjusting the number or weight ratio of sub-peaks, which has no physical meaning. Due to the lack of a unified standard for mutual comparison, it is difficult to reflect the small wavenumber changes caused by stress in the fitting results, which makes the multimodal fitting method insensitive and inaccurate for use in stress analysis.

The kernel of the barycenter method is to use the barycentric peak instead of the characteristic peak for data processing. In the experiment, environmental factors had a great impact on the THz time-domain spectral signals due to the long collection time (100 ps), which led to a low signal-to-noise ratio. In the barycenter method, the integral operation of the peak area for calculating the barycentric peak can accumulate and eliminate random noise, further identifying the position of the spectral position. In addition, the barycenter method is more suitable for the data processing of spectra with poor symmetry, because the area integral operation can avoid the interference of spectral asymmetry in data processing. As shown in Figure 5c and Table 1, the stress optical coefficient c_1 obtained by the barycenter method was close to that of the multimodal fitting method, and the time delay shows a linear decrease with compressive stress, which verifies the feasibility of the barycenter method. In addition, the linear GOF of the barycenter method is the highest among the three methods, indicating that it is more suitable for processing the THz spectra.

Based on a reflection-type THz-TDS system and the barycenter method, the stress optical coefficient c_1 of the TBC specimen is 0.3180 ± 0.0091/GPa. As shown in Figure 5c, the refractive index of the TBC specimen under different compressive loadings has an excellent linear relationship with the stress, which proves the accuracy of the stress optical coefficient. In addition, the calibrated stress optical coefficient in this work is numerically different from that detected by Schemmel et al. under tension loading (0.1362 ± 0.0155/GPa) [32], which is caused by the differences in preparation methods and stress states of the TBC specimens. The TC layer in a TBC is generally a porous and brittle ceramic material. Different stress states (tension or compression), preparation methods and production batches have great influence on the mechanical properties of the TC layer, including its stress optical coefficient. Compared to the results of Schemmel et al., the measured stress optical coefficient in this work was the that of the TBC specimen under compression loadings, instead of tension loadings. In addition, the raw materials and processing methods of the TBC specimen used in this work were different from those adopted by Schemmel et al. Thus, the stress optical coefficient was numerically different from that measured by Schemmel et al. In most cases, the stress state of the TBC top coating is compressive stress. Hence, for the internal stress analysis of the TBC top coating, it is necessary to measure the stress optical coefficient of the TBC top coating under compression.

5. Conclusions

In this work, the stress optical coefficient of the TBC specimen was calibrated for the stress analyses in the ceramic layer of TBC based on THz-TDS. The THz time-domain spectra of TBC specimens under different uniaxial compression loadings were obtained by using a reflection-type THz-TDS system and were processed by the unimodal fitting, multimodal fitting and barycenter methods, respectively. By comparing the results of these three methods, the barycenter method showed higher accuracy and stability than the other two methods. The stress optical coefficient c_1 of the ceramic specimen obtained in this work, 0.3180 ± 0.0091/GPa, is quite different from those detected under tensile stress. This result lays a foundation for using THz-TDS to analyze the internal stress of the top coating of actual TBC structures.

Author Contributions: Conceptualization, W.Q.; methodology, Y.Z., N.L. and Z.W. (Zong Wang); formal analysis, Y.Z., N.L. and Z.W. (Zong Wang); investigation, Y.Z., N.L. and Z.W. (Zong Wang); resources, W.Q.; data curation, Y.Z., N.L. and Z.W. (Zong Wang); writing—original draft preparation, Z.W. (Zong Wang) and Y.Z.; writing—review and editing, Z.W. (Zhiyong Wang), W.Q., Y.Z. and Z.W. (Zong Wang); visualization, Z.W. (Zhiyong Wang), W.Q., Y.Z. and Z.W. (Zong Wang); supervision, Z.W. (Zhiyong Wang) and W.Q.; project administration, W.Q.; funding acquisition, W.Q. All authors have read and agreed to the published version of the manuscript.

Funding: This work was supported by the National Natural Science Foundation of China (Grant Nos.12021002, 11827802, and 11890680).

References

1. Wang, D.R.; Fan, X.L.; Sun, Y.L. The stresses and cracks in thermal barrier coating system: A review. *Chin. J. Solid Mech.* **2016**, *37*, 477–517. [CrossRef]
2. Clarke, D.R.; Oechsner, M.; Padture, N.P. Thermal-barrier coatings for more efficient gas-turbine engines. *MRS Bull.* **2012**, *37*, 891–899. [CrossRef]
3. Schulz, U.; Leyens, C.; Fritscher, K. Some recent trends in research and technology of advanced thermal barrier coatings. *Aerosp. Sci. Technol.* **2003**, *7*, 73–80. [CrossRef]
4. Padture, N.P. Advanced structural ceramics in aerospace propulsion. *Nat. Mater.* **2016**, *15*, 804–809. [CrossRef] [PubMed]
5. Eldridge, J.I.; Bencic, T.J.; Spuckler, C.M. Delamination-indicating thermal barrier coatings using YSZ: Eu sublayers. *J. Am. Ceram. Soc.* **2006**, *89*, 3246–3251. [CrossRef]
6. Guo, H.B.; Kuroda, S.; Murakami, H. Microstructures and properties of plasma-sprayed segmented thermal barrier coatings. *J. Am. Ceram. Soc.* **2006**, *89*, 1432–1439. [CrossRef]
7. Nychka, A.; Clarke, D.R. Damage quantification in TBCs by photo-stimulated luminescence spectroscopy. *Surf. Coat. Technol.* **2001**, *146*, 110–116. [CrossRef]
8. Jiang, P.; Fan, X.L.; Sun, Y.L. Competition mechanism of interfacial cracks in thermal barrier coating system. *Mater. Des.* **2017**, *132*, 559–566. [CrossRef]
9. Jiang, P.; Fan, X.L.; Sun, Y.L. Bending-driven failure mechanism and modelling of double-ceramic-layer thermal barrier coating system. *Int. J. Solids Struct.* **2017**, *130*, 11–20. [CrossRef]
10. Miller, R.A. Oxidation-based model for thermal barrier coating life. *J. Am. Ceram. Soc.* **2010**, *67*, 517–521. [CrossRef]
11. Widjaja, S.; Limarga, A.M.; Yip, T.H. Modeling of residual stresses in a plasma-sprayed zirconia/alumina functionally graded-thermal barrier coating. *Thin Solid Film.* **2003**, *434*, 216–227. [CrossRef]
12. Kuroda, S.; Fukushima, T.; Kitahara, S. Generation mechanisms of residual stresses in plasma-sprayed coatings. *Vacuum* **1990**, *41*, 1297–1299. [CrossRef]
13. Matejícek, J.; Sampath, S.; Dubsky, J. X-ray residual stress measurement in metallic and ceramic plasma sprayed coatings. *J. Therm. Spray Technol.* **1998**, *7*, 489–496. [CrossRef]
14. Totemeier, T.C.; Wright, J.K. Residual stress determination in thermally sprayed coatings—A comparison of curvature models and X-ray techniques. *Surf. Coat. Technol.* **2006**, *200*, 3955–3962. [CrossRef]
15. Wang, L.; Wang, Y.; Sun, X.G.; He, J.Q.; Pan, Z.Y.; Wang, C.H. Microstructure and indentation mechanical properties of plasma sprayed nano-bimodal and conventional ZrO_2-8 wt.% Y_2O_3 thermal barrier coatings. *Vacuum* **2012**, *86*, 1174–1185. [CrossRef]
16. Jordan, D.W.; Faber, K.T. X-ray residual stress analysis of a ceramic thermal barrier coating undergoing thermal cycling. *Thin Solid Film.* **1993**, *235*, 137–141. [CrossRef]
17. Mao, W.G.; Chen, Q.; Dai, C.Y.; Yang, L.; Zhou, Y.C.; Lu, C. Effects of piezo-spectroscopic coefficients of 8 wt.% Y_2O_3 stabilized ZrO_2 on residual stress measurement of thermal barrier coatings by Raman spectroscopy. *Surf. Coat. Technol.* **2010**, *204*, 3573–3577. [CrossRef]
18. Loechelt, G.H.; Cave, N.G.; Menendez, J. Measuring the tensor nature of stress in silicon using polarized off-axis Raman spectroscopy. *Appl. Phys. Lett.* **1995**, *66*, 3639–3641. [CrossRef]
19. Fu, D.H.; He, X.Y.; Ma, L.L.; Xing, H.D.; Meng, T.; Chang, Y.; Qiu, W. The 2-axis stress component decoupling of {100} c-Si by using oblique backscattering micro-Raman spectroscopy. *Sci. China Phys. Mech.* **2020**, *63*, 10460. [CrossRef]
20. Zhao, Y.; Ma, C.L.; Huang, F.X.; Wang, C.J.; Zhao, S.M.; Cui, X.Q.; Li, F.F. Residual stress inspection by Eu^{3+} photoluminescence piezo-spectroscopy: An application in thermal barrier coatings. *J. Appl. Phys.* **2013**, *114*, 073502. [CrossRef]
21. Zhao, S.M.; Ren, Z.M.; Zhao, Y.; Xu, J.Y.; Zou, B.L.; Hui, Y.; Zhu, L.; Zhou, X.; Cao, X.Q. The application of Eu^{3+} photoluminescence piezo-spectroscopy in the $LaMgAl_{11}O_{19}$/8YSZ:Eu double-ceramic-layer coating system. *J. Eur. Ceram. Soc.* **2015**, *35*, 249–257. [CrossRef]
22. Jiang, P.; Fan, X.L.; Sun, Y.L.; Wang, H.T.; Su, L.C.; Wang, T.J. Thermal-cycle dependent residual stress within the crack-susceptible zone in thermal barrier coating system. *J. Am. Ceram. Soc.* **2018**, *101*, 4256–4261. [CrossRef]
23. Fan, W.H.; Burnett, A.; Upadhya, P.C. Far-Infrared spectroscopic characterization of explosives for security applications using broadband terahertz time-domain spectroscopy. *Appl. Spectrosc.* **2007**, *61*, 638–643. [CrossRef]
24. Kojima, S.; Tsumura, N.; Kitahara, H. Terahertz time domain spectroscopy of phonon-polaritons in ferroelectric lithium niobate crystals. *Jpn. J. Appl. Phys.* **2002**, *41*, 7033–7037. [CrossRef]
25. Cogdill, R.P.; Forcht, R.N.; Shen, Y. Comparison of terahertz pulse imaging and near-infrared spectroscopy for rapid, non-destructive analysis of tablet coating thickness and uniformity. *Pharm. Innov.* **2007**, *2*, 29–36. [CrossRef]
26. Zhang, X.N.; Chen, J.; Zhou, Z.K. THz time-domain spectroscopy technology. *Laser Optoelectron. Prog.* **2005**, *42*, 35–38. (In Chinese)
27. Zhong, S.C. Progress in terahertz nondestructive testing: A review. *Front. Mech. Eng.* **2019**, *14*, 273–281. [CrossRef]
28. Song, W.; Li, L.A.; Wang, Z.Y. Experimental verification of the uniaxial stress-optic law in the terahertz frequency regime. *Opt. Lasers Eng.* **2007**, *52*, 174–177. [CrossRef]
29. Shuai, S.; Zhang, G.; Shi, W. Optically pumped gas terahertz fiber laser based on gold-coated quartz hollow-core fiber. *Appl. Opt.* **2019**, *58*, 2828–2831. [CrossRef]

30. Chen, C.C.; Lee, D.J.; Pollock, T. Pulsed-terahertz reflectometry for health monitoring of ceramic thermal barrier coatings. *Opt. Express* **2010**, *18*, 3477–3486. [CrossRef]
31. Watanabe, M.; Kuroda, S.; Yamawaki, H. Terahertz dielectric properties of plasma-sprayed thermal-barrier coatings. *Surf. Coat. Technol.* **2011**, *205*, 4620–4626. [CrossRef]
32. Schemmel, P.; Diederich, G.; Moore, A.J. Direct stress optic coefficients for YTZP ceramic and PTFE at GHz frequencies. *Opt. Express* **2016**, *24*, 8110–8119. [CrossRef] [PubMed]
33. Schemmel, P.; Diederich, G.; Moore, A.J. Measurement of direct strain optic coefficient of YSZ thermal barrier coatings at GHz frequencies. *Opt. Express* **2017**, *24*, 19968–19980. [CrossRef] [PubMed]
34. Fukuchi, T.; Fuse, N.; Okada, M. Measurement of refractive index and thickness of topcoat of thermal barrier coating by reflection measurement of terahertz waves. *Electr. Commun. Jpn.* **2013**, *96*, 37–45. [CrossRef]

Article

Experiment Study on Topological Characteristics of Sandstone Coating by Micro CT

Feng Gao [1,2], Yuhao Hu [1,2], Guannan Liu [1,2,3,*] and Yugui Yang [1,2]

[1] State Key Laboratory for Geomechanics and Deep Underground Engineering,
China University of Mining and Technology, Xuzhou 221116, China; fgao@cumt.edu.cn (F.G.);
TS20030008A31@cumt.edu.cn (Y.H.); yyg323@163.com (Y.Y.)

[2] Mechanics and Civil Engineering Institute, China University of Mining and Technology,
Xuzhou 221116, China

[3] Laboratory of Mine Cooling and Coal-Heat Integrated Exploitation,
China University of Mining and Technology, Xuzhou 221116, China

* Correspondence: guannanliu@126.com

Received: 27 October 2020; Accepted: 19 November 2020; Published: 24 November 2020

Abstract: The pore structure is an important factor of tunnel coating failure, cracking and water leakage. Some investigations on the statistical law of pores and pore networks have been conducted, but little quantitative analysis is observed on topology structure of the pore network, and even the pore structure of sandstone is complex and cross-scale distributed. Therefore, it is of theoretical and engineering significance to quantitatively characterize the connectivity of the pore network in sandstone. This study proposes a new complex network theory to analyze the three-dimensional nature of pore network structure in sandstone. The topological network structure, such as clustering degree, average path length and the module, which cannot be analyzed by traditional coordination number and fractal dimension methods, is analyzed. Numerical simulation results show that a scale-free network model is more suitable for describing the sandstone pore network than random models. The pore network of sandstone has good uniformity. The connectivity of sandstone pore networks has great potential for permeability enhancement. Therefore, this new method provides a way to deeply understand the pore connectivity characteristics of sandstone and to explore the distribution of crack grids in the arch of tunnel coatings.

Keywords: complex network theory; sandstone; network topology; seepage; distribution of degree

1. Introduction

An important problem in tunnels is how to effectively reduce the coating leakage, so the connectivity of rock pore structure and the waterproof materials in the coating are the key issues [1–3]. The structure of the rock pore network is the result of the interaction among its physical and chemical properties, and external stress environments. Rock has large numbers of pores, whose shapes and connected structures are complex [4–6]. How to comprehensively and quantitatively characterize the three-dimensional pore structure and explore the topological characteristics of effective seepage networks is of important theoretical significance and engineering application value [7]. This can help us to reveal the seepage mechanism and effectively reduce the leakage in tunnels [8–11].

In recent years, a lot of studies have been conducted on the statistical characteristics and evolution model of rock pore structure [12–16]. Blunt and Raeini [4,17] developed a generalized network extraction method for three-dimensional digital core to reduce the uncertainties of modeling. Using a focused ion beam scanning electron microscope (FIB-SEM), Shaina et al. [18] obtained a microscale shale digital core and compared it with the conventional scanning electron microscopy. The porosity, organic content and pore connectivity were studied. Hajizadeh [19] took the CT(computerized tomography) scan image data of sandstone as a sample of a statistical method. They coupled the continuous two-dimensional multi-point statistical simulation with the multi-scale data extraction program, and proposed a random reconstruction technique for porous media. As the traditional Darcy law is no longer suitable for dense rock, Civan et al. [20] used the modified Darcy law to describe the gas migration in dense shale. The apparent permeability is a function of intrinsic permeability and porosity. Zhang et al. [21] considered the basic concept of a pore network model and proposed a modeling method based on pore volume and throat searching. They combined the pore network model with the percolation theory to calculate the permeability of different models. Ju et al. [22] used fractal control function to describe the complex morphology of rock pore structure, and proposed a fractal reconstruction model of rock pore structure with an improved simulated annealing algorithm. From the pore number per unit volume, porosity, average pore size, pore volume, minimum maximum pore radius and minimum average pore radius, Gong et al. [23] analyzed the pore radius and average pore shape factor of the eight pore structure parameters of sandstone pore structure characteristics. Usually, nuclear magnetic resonance and CT images are used to extract the three-dimensional structure of rock pores, and the quantitative results describe the pore structure of rocks by specific surface area, surface roughness and fractal dimension. However, the selection of seepage path and pore community structure cannot be quantified.

As the main component of spraying materials for tunnel coating materials, the pore structure characteristics of sandstone play a key role in the deformation and permeability of tunnel coatings. In this study, the topological characteristics and correlation properties of pore networks are analyzed by a new method—complex network theory. Three new network parameters of sandstone with different porosity are analyzed: (1) degree distribution; (2) modularity; (3) degree of pore aggregation. This proposed method can quantitatively analyze the structure parameters which cannot be quantified by traditional connectivity analysis and fractal dimension analysis. This provides a new way to further understand the mechanism of the pore network structure characteristics to the macroscopic permeability. A new method for sorting the importance of sandstone pores, called "eigenvector centrality", is given and provides a basis for revealing the microscopic mechanism of porous rock seepage.

2. Basic Information of Sandstone Pore Network

2.1. Sandstone Network Extraction

This study uses the experimental data of a sandstone core which were obtained by Professor Blunt of Imperial College London. Micro CT produced by the German Phoenix company was used to analyze the image, which is equipped with a 1 m focus system with a view of 512×512 pixels with a 0.2 m 16-bit detector. The tomographic images were generated by X-ray radiation, and then transferred into binary images according to the porosity of each sample. Four sandstone samples with porosities of 14.1%, 16.9%, 17.1% and 24.6% are used in our study. The sandstone networks are extracted from the 3D digital core. The network scale is 3 mm by 3 mm by 3 mm, which was chosen from the cubes inside the four samples.

More than 4000 nodes were extracted from each core. In order to analyze the effective seepage network, the following two corrections are made in this study. Firstly, isolated pores, which are not connected with the main network, are ignored. This is because the isolated pores do not influence the seepage process. They were removed during the pore network construction of sandstone. Then, ring-shaped throats are deleted. The three-dimensional sandstone pore networks of the four digital cores are obtained based on the above calculations and shown in Figure 1 [24].

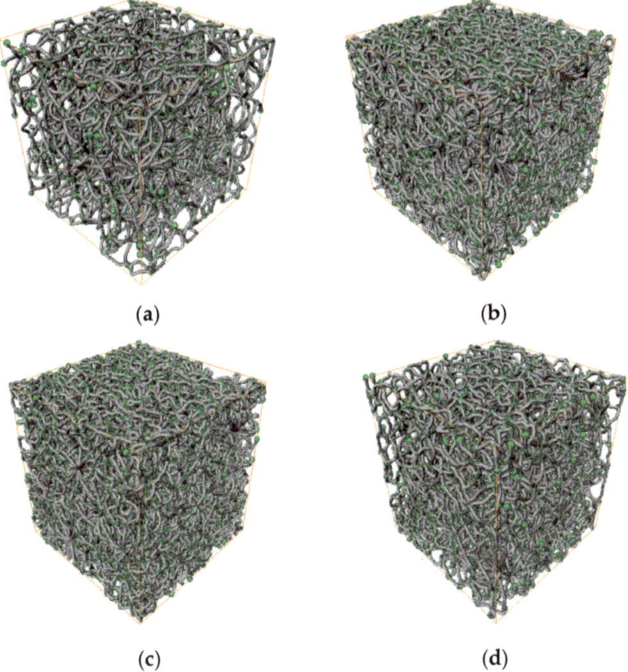

(a)   (b)

(c)   (d)

Figure 1. Sandstone three-dimensional pore network (bounding box volume: 3 mm by 3 mm by 3 mm. The pore network is extracted by the skeleton algorithm in the digital rock reconstructed from the X-ray CT scanner): (**a**) 14.1%; (**b**) 16.9%; (**c**) 17.1%; (**d**) 24.6%.

2.2. Basic Parameters of Sandstone Networks in Complex Network Theory

The rock pore structure is commonly described in quantitative analysis by porosity, pore diameter, pore throat radius, specific surface area, Euler number and coordination number [25–27]. However, the network topological properties are yet to be found. This study uses complex network theory to study the quantitative characterization of the sandstone pore structure. The basic information of the pore network structure of sandstone is firstly analyzed. The analysis results are listed in Table 1.

Table 1. Basic parameters of sandstone networks.

Porosity ϕ	Node Number N	Edge Number M	Average Degree $\langle k \rangle$	Network Diameter D	Average Path Length L	Network Density ρ
14.1%	1717	2824	3.289	25	10.151	0.002
16.9%	8301	14,381	3.465	46	16.469	0
17.1%	8542	12,432	2.962	50	19.428	0
24.6%	1945	4697	4.83	22	8.466	0.002

This table shows that the pore network properties of sandstone are mainly related to network size (nodal number) rather than sandstone porosity. When the scale of the network increases, the diameter of the network and the average path length grow correspondingly. In complex network theory, the degree k of node i is used to express the connectivity of the node. It is defined as the number of other nodes connected with this node. The average degree of all nodes in the network is called the average degree of the network (denoted as $\langle k \rangle$). Because the contribution of each pair of edges is 2, the average degree of the network is:

$$\langle k \rangle = \frac{1}{N}\sum_{i=1}^{N} k_i = \frac{2M}{N} \tag{1}$$

where M represents the number of edges of the network; N represents the number of nodes of the network.

The network diameter D and the average path length L are also used to measure the network transmission efficiency. The maximum value of the shortest path length d between any two nodes in the network is called the network diameter:

$$D = \max_{i,j}(d_{ij}) \tag{2}$$

The distance between any two nodes in a network is defined as the number of edges of the shortest path between the two nodes. The average path length L is defined as the average distance between all nodes in a network:

$$L = \frac{2}{N(N-1)}\sum_{i=1}^{N-1}\sum_{j=i+1}^{N} d_{ij} \tag{3}$$

Further, network density is used to describe the degree of denseness between nodes in a network. It is defined as the ratio of the actual number of edges to the upper limit of the number of edges in the network:

$$\rho = \frac{2M}{N(N-1)} \tag{4}$$

According to Table 1, the network diameter and average path length of the sandstone pore network are large, but the network density is small. So, the network structure connectivity of the sandstone pore network is much smaller than that of the global coupling network (any two nodes are connected).

3. Sandstone Network Structure Analysis in Complex Network Theory

In order to further analyze the microstructure characteristics of the sandstone pore network, the degree distributions of nodes in the network are analyzed. Figure 2 is the degree distribution of three-dimensional pore networks of the four sandstone cores. The degree distributions of the sandstone pore network are single-peaked curves, but the sandstone pore networks have different characteristic values which are variable with porosity. When the porosities are 14.1%, 16.9% and 17.1%, the nodes have a moderate proportion which has a degree of 2. When the porosity is 24.6%, the highest proportion of nodes is the nodes whose degree is 4. When the degree exceeds these values, the number of nodes decreases abruptly.

According to the density of the number of edges among nodes, a network can be divided into different groups called communities. For the community structure identification in the network, different algorithms have been put forward, such as the hierarchical clustering method and shortest path betweenness method. The shortest path betweenness method can not only give the results of community structure division, but also give the evaluation of this division. This is called modularity.

A symmetric matrix $E = (e_{ij})$ is defined, in which the element e_{ij} represents the proportion of the edges of the i community and the j community in all edges of the network. The sum of the elements on the diagonal is $TrE = \sum_{i} e_{ij}$, which represents the proportion of the edges between the nodes in a community within all edges. The sum of the elements in each row is $a_i = \sum_{j} e_{ij}$, indicating the proportion of the edges connected to the nodes in the i community in all edges. On this basis, the definition of modularity is given:

$$Q = \sum_{i} \left(e_{ij} - a_i^2 \right) \tag{5}$$

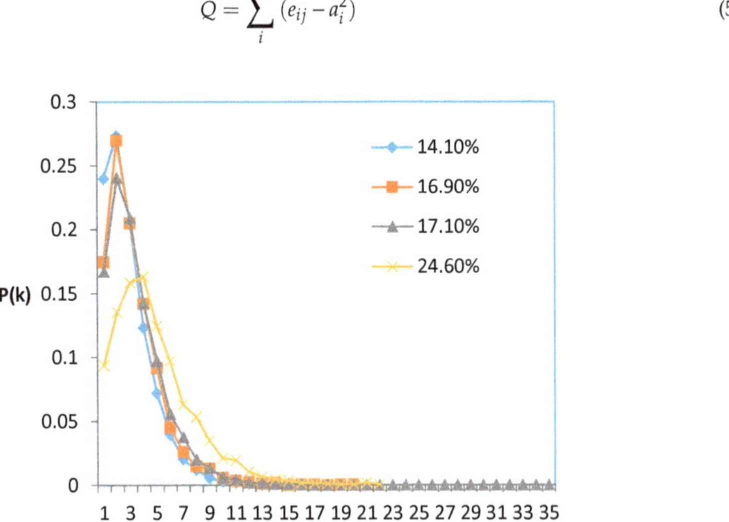

Figure 2. Degree distribution of sandstone three-dimensional pore network.

Modularity measures the difference between the proportion of the inner edge of the community and the proportion among communities. When the proportion of the inner edge of the community is equal to that of the edge among the community, there is $Q = 0$, which means that there is no obvious difference of the proportion between the inside and outside edges of the community, and there is no obvious community. On the contrary, if $Q \rightarrow 1$, the proportion of the inner side of the community is much higher, indicating that the division of the community is very obvious. Usually, when $Q \geq 0.3$, there is an obvious community structure.

The modularity of each network is calculated and shown in Figure 3. When the porosity of the sandstone is 17.1%, its pore network module is approximately equal to 0.919 and the pore size of different sandstone pore networks is close to 1. This shows that the sandstone pore network has a good community structure. The nodes are grouped according to different societies, and the results are shown in Figure 4, where the same community nodes are of the same color and size.

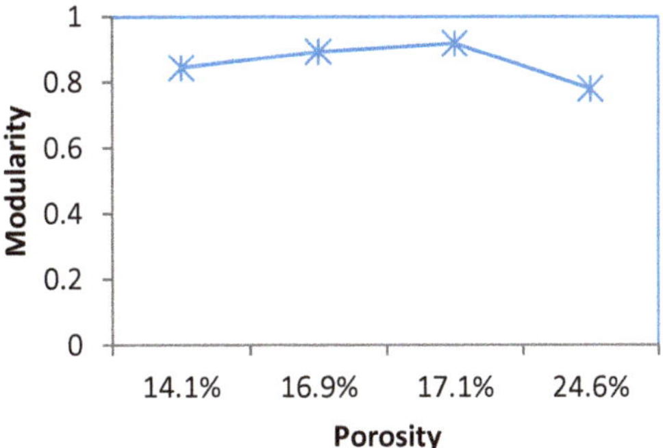

Figure 3. Sandstone pore network modularity.

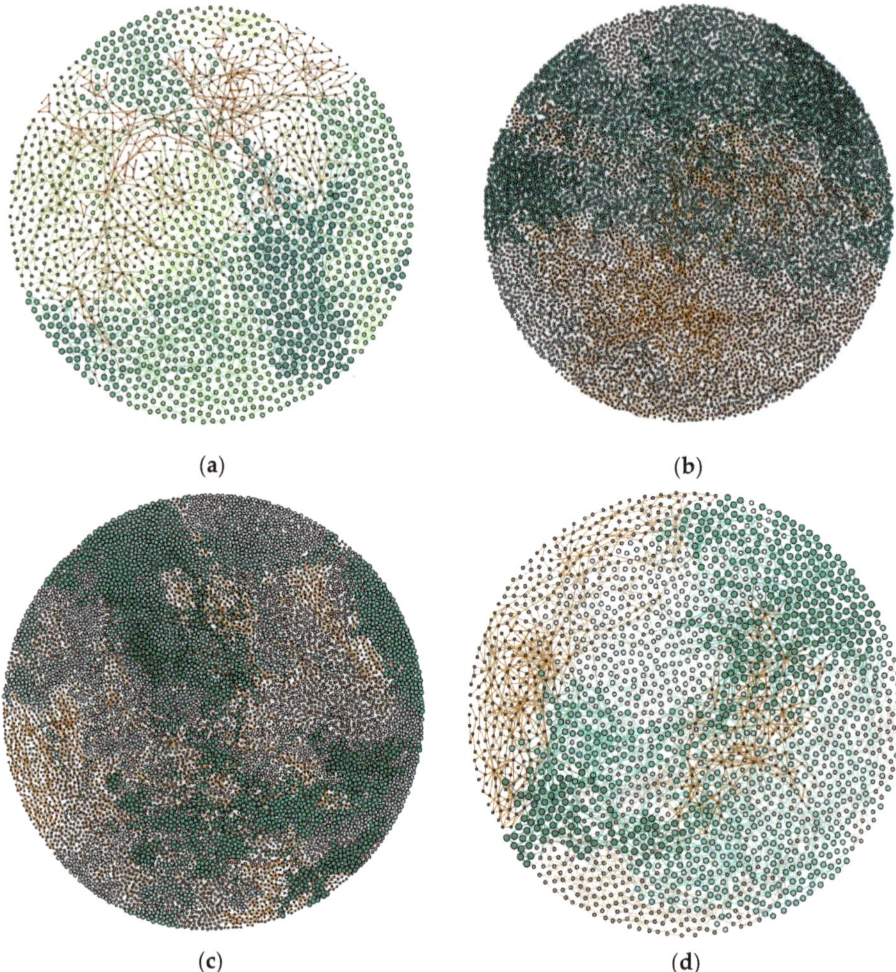

Figure 4. Association structure of sandstone pore network: (**a**) 14.1%; (**b**) 16.9%; (**c**) 17.1%; (**d**) 24.6%.

As shown in Figure 5, when the porosity is 14.1% and 24.6%, the number of nodes in the pore network is less, and the number of nodes in each community is less than 200, which is about 1/10 of the total number of nodes. When the porosity is 16.9% and 17.1%, with the increase in network size, the number of nodes in a single community increases obviously. In the four samples, when the porosity is 16.9% and modularity class is 110, the maximum number of nodes in the network community is about 375, less than 1/20 of the total number of nodes. These show that sandstone pore network distribution is uniform, and there is no phenomenon that some communities play a leading role in the whole sandstone network.

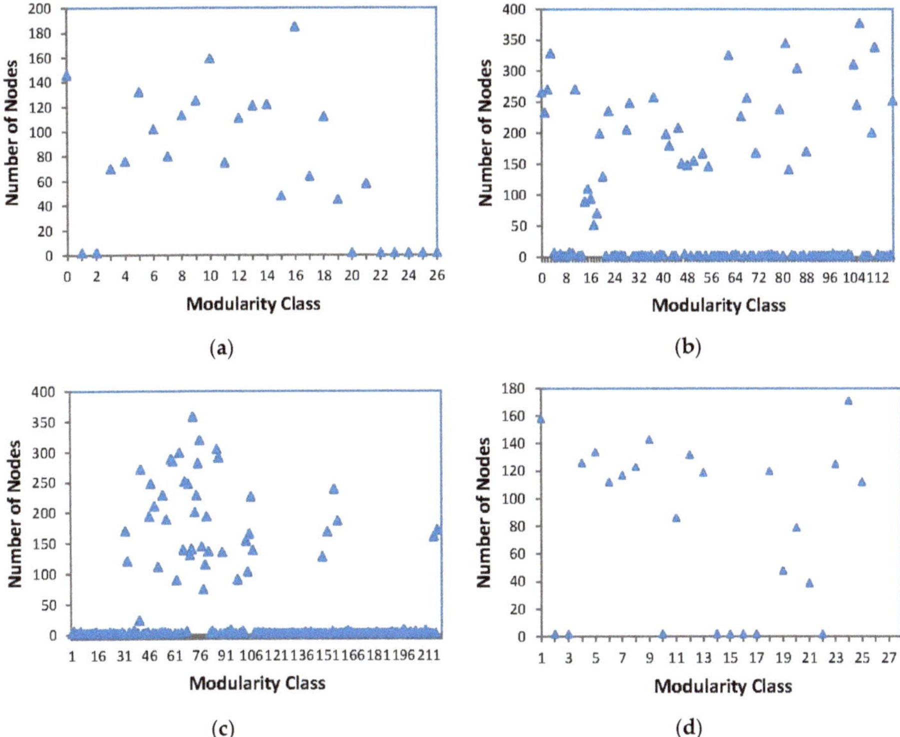

Figure 5. Distribution of node numbers in network community of sandstone: (**a**) 14.1%; (**b**) 16.9%; (**c**) 17.1%; (**d**) 24.6%.

For the further measurement of the sandstone pore network transmission mechanism of the recovery enhancement, Liu [15] has carried out quantitative analysis for the pore structure by fractal dimension. However, the discussion about the microscopic connectivity characteristic of the sandstone pore network is sparse. This study chooses the clustering coefficient related to transmission efficiency to analyze the sandstone pore network. The clustering coefficient is the coefficient that indicates the degree of node aggregation in a graph. In the network of reality, the higher the clustering coefficient, the more nodes tend to be closely connected. In the network, if the node i is connected to the node $i+1$ and the node $i+1$ is connected to the node $i+2$, the node $i+2$ may be connected to i. This phenomenon reflects the connection density between nodes. In an undirected network, the clustering coefficients can be expressed as:

$$C = \frac{n}{C_k^2} = \frac{2n}{k(k-1)} \tag{6}$$

where n represents the number of edges that are connected among all adjacent nodes of the node i. The clustering coefficient distribution of the sandstone pore network is shown in Figure 6. This figure shows that the clustering coefficients of different sandstone pore networks are mainly distributed between 0 and 0.5. The nodes with the clustering coefficient of 0 have the largest proportion, and the number of nodes decreases as the clustering coefficient increases. It shows that the sandstone pore network has less connectivity than the global coupling network whose clustering coefficient is 1. According to Figure 2, the main reason is that the main pore connected to the sandstone pore is the peripheral pore node. In general, as the distance increases, the probability of connection between two nodes becomes weaker.

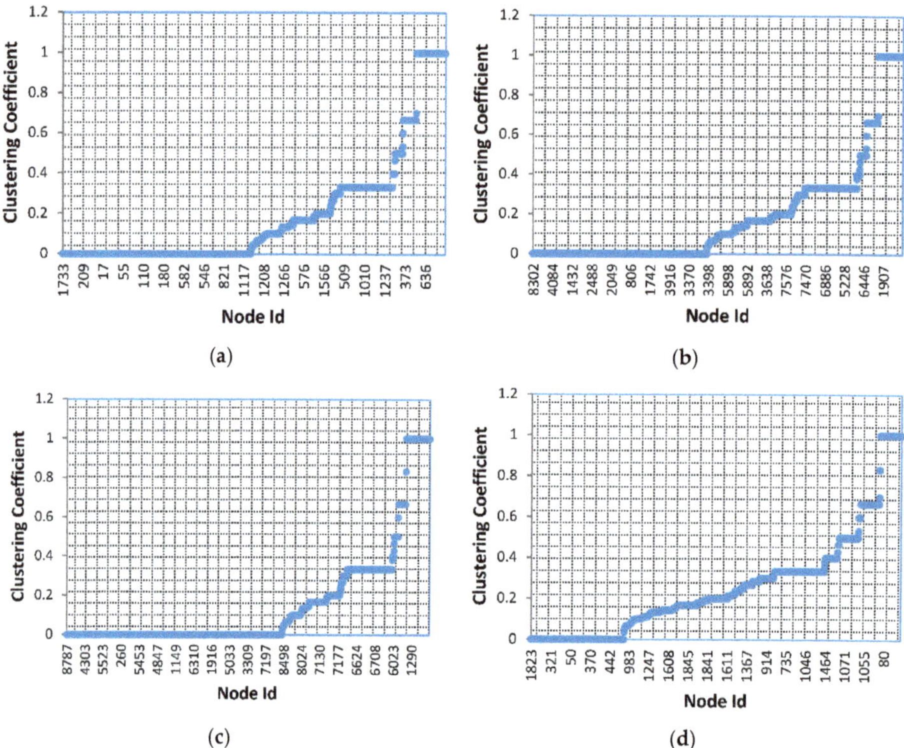

Figure 6. Clustering coefficient distribution of sandstone three-dimensional pore network: (**a**) 14.1%; (**b**) 16.9%; (**c**) 17.1%; (**d**) 24.6%.

The degree of aggregation of the entire network can be represented by the average value of the clustering coefficients of all nodes, as shown in Figure 7. The networks with higher average clustering coefficients have smaller average distances in different nodes. The average clustering coefficient is the largest when the porosity is 24.60% in the sampling sandstone and, as shown in Table 1, the average path length is the minimum. That is, the network has the best connectivity among the four sandstone samples.

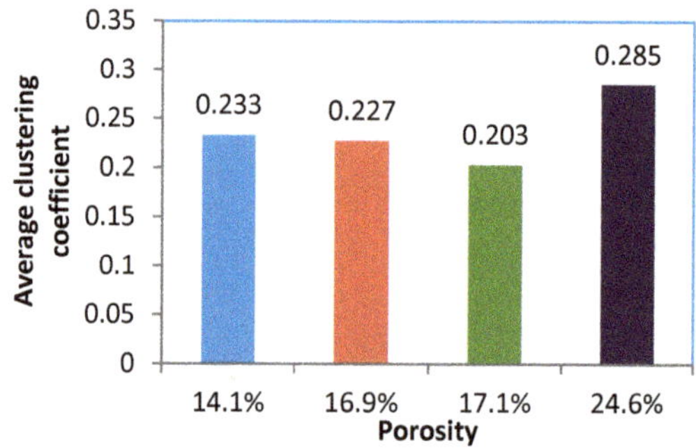

Figure 7. Average clustering coefficient of sandstone pore network.

For the data in Figure 7, a simple one-way ANOVA is performed, as shown in Table 2. We can see that F is less than F crit and P is greater than 0.05, so the porosity and average clustering coefficient have statistical significance.

Table 2. One-way ANOVA.

Differences between the Sources	SS	df	MS	F	*p*-Value	F Crit
between groups	0.006105	1	0.006105	3.798787	0.09919	5.987378
within the same group	0.009643	6	0.001607	-	-	-
sum	0.015748	7	-	-	-	-

In order to analyze the importance of individual nodes in sandstone pore networks, the eigenvector centrality is used to compute the network nodes. Make x_i the index value of node i, and A_{ij} the adjacency matrix of the network. For node i, the centrality index is proportional to the exponential sum of all nodes connected to it. That is:

$$x_i = \frac{1}{\lambda} \sum_{j \in M_i} x_j = \frac{1}{\lambda} \sum_{j=1}^{N} A_{i,j} x_j \tag{7}$$

where M_i is the set of nodes connected to node i. λ is constant.

The centrality of the eigenvector is shown in Figure 8. The abscissa in the figure is the node Id. When the porosity of sandstone is between 14.1% and 24.6%, the eigenvector centrality of most nodes is less than 0.4. Only a small number of nodes have high centrality.

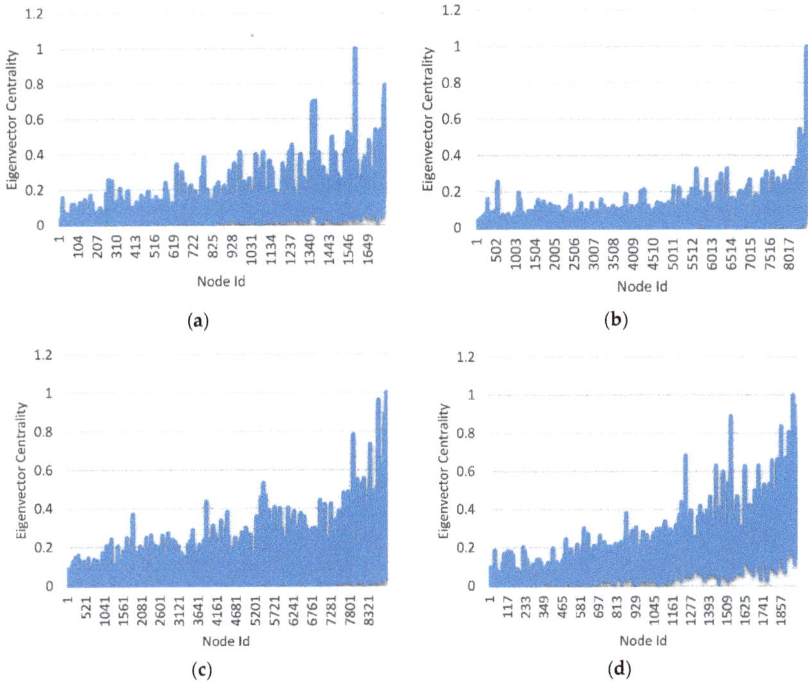

Figure 8. Eigenvector centrality distribution of sandstone network nodes: (**a**) 14.1%; (**b**) 16.9%; (**c**) 17.1%; (**d**) 24.6%.

To facilitate intuitive understanding, the Fruchterman–Reingold algorithm is used to distribute the pore nodes in Figure 1 uniformly on a circular surface. The node centrality is sorted in Figure 9, where the node size represents the node centrality, and the number represents the node Id.

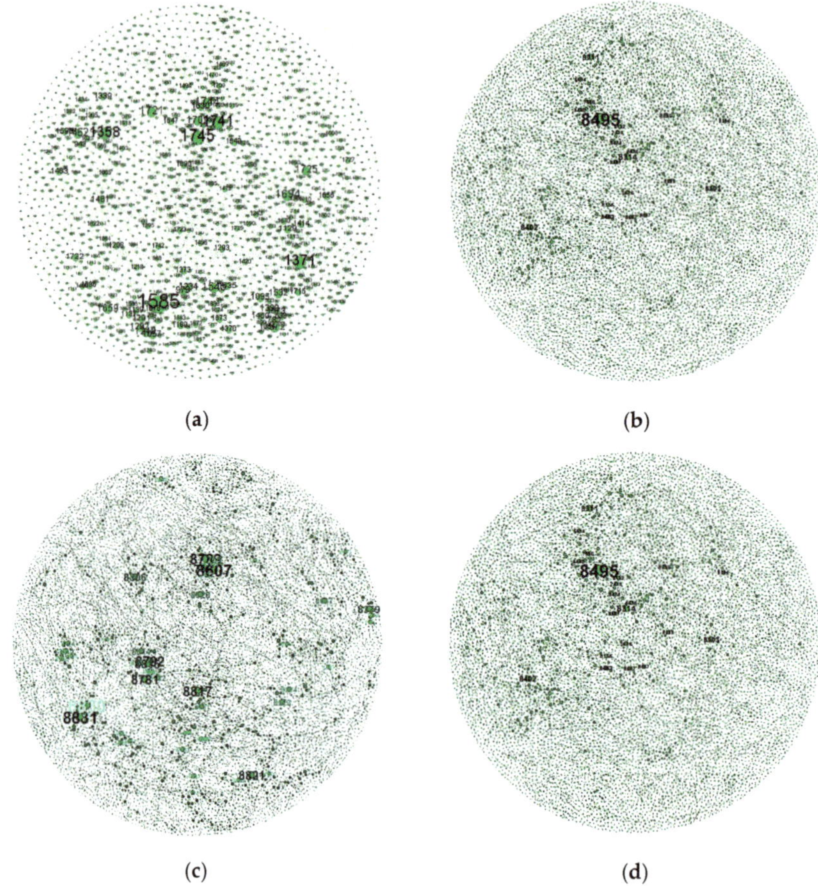

<table>
<tr><td align="center">(a)</td><td align="center">(b)</td></tr>
<tr><td align="center">(c)</td><td align="center">(d)</td></tr>
</table>

Figure 9. Centrality sorting of sandstone pore nodes: (**a**) 14.1%; (**b**) 16.9%; (**c**) 17.1%; (**d**) 24.6%.

4. Conclusions

In this study, complex network theory was applied to investigate the network characteristics of the pore structure, which cannot be quantitatively analyzed by traditional methods such as Euler number and fractal dimension. Three network structure parameters, including degree distribution, clustering coefficient and modularity, were taken into account. A new eigenvector centrality method was used to sort the importance of the nodes in the percolation network. Based on these studies, following conclusions can be drawn:

(1) The results show that a scale-free model is more suitable for the degree distribution of the sandstone seepage network. Without considering pore volume and pore throat volume weight, the overall permeability of the network may be determined by a small number of pores with large degree (number of connections with other pores).

(2) The modular degree Q of the pore network of different porosity sandstone tends to 1, which means that the pore network of sandstone has an obvious modularized structure. According to the

network topology, the sandstone network can be divided into several relatively independent communities. The associations within the community are closely connected, and have strong permeability, while the connection between communities is less, which inhibits the overall permeability of sandstone.

(3) The aggregation coefficients of different sandstone pore networks are mainly distributed between 0 and 0.5. The connectivity of the sandstone pore network is far less than that of fully coupled networks with all nodes connected. That is, the sandstone has a strong antireflection potential.

Author Contributions: Data curation, F.G.; Formal analysis, Y.H.; Funding acquisition, F.G.; Investigation, Y.H.; Methodology, Y.Y.; Software, G.L.; Writing–original draft, F.G.; Writing–review and editing, G.L. All authors have read and agreed to the published version of the manuscript.

Funding: This work was supported by the Fundamental Research Funds for the Central Universities (Grant No. 2020ZDPYMS02).

Conflicts of Interest: The authors declare no conflict of interest.

References

1. Chen, P.; Chen, S.; Peng, J.; Gao, F.; Liu, H. The interface behavior of a thin film bonded imperfectly to a finite thickness gradient substrate. *Eng. Fract. Mech.* **2019**, *217*, 106529. [CrossRef]
2. Lou, L.; Chen, P.; Wang, Z.; Zhang, S.; Gao, F. Cohesive energy measurement of van der Waals heterostructures by the shaft loaded blister test. *Extreme Mech. Lett.* **2020**, *41*, 100987. [CrossRef]
3. Chen, P.; Chen, S.H.; Liu, H.; Peng, J.; Gao, F. The interface behavior of multiple piezoelectric films attaching to a finite-thickness gradient substrate. *J. Appl. Mech. Trans. ASME* **2020**, *87*, 011003. [CrossRef]
4. Blunt, M.J.; Bijeljic, B.; Hu, D.; Gharbi, O.; Iglauer, S.; Mostaghimi, P.; Paluszny, A. ChristopherPentlan Pore-scale imaging and modelling. *Adv. Water Resour.* **2013**, *51*, 197–216. [CrossRef]
5. Arand, F.; Hesser, J. Accurate and efficient maximal ball algorithm for pore network extraction. *Comput. Geosci.* **2017**, *101*, 28–37. [CrossRef]
6. Bauer, B.; Cai, X.; Peth, S.; Schladitz, K.; Steidl, G. Variational-based segmentation of bio-pores in tomographic images. *Comput. Geosci.* **2017**, *98*, 1–8. [CrossRef]
7. Chen, P.; Peng, J.; Chen, Z.; Peng, G. On the interfacial behavior of a piezoelectric actuator bonded to a homogeneous half plane with an arbitrarily varying graded coating. *Eng. Fract. Mech.* **2019**, *220*, 106645. [CrossRef]
8. Liu, G.; Yu, B.; Gao, F.; Ye, D.; Yue, F. Analysis of permeability evolution characteristics based on dual fractal coupling model for coal seam. *Fractals* **2020**, in press. [CrossRef]
9. Li, K.; Kong, S.; Xia, P.; Wang, X. Microstructural characterisation of organic matter pores in coal-measure shale. *Adv. Geo-Energy Res.* **2020**, *4*, 372–391. [CrossRef]
10. Li, Z.; Duan, Y.; Fang, Q.; Wei, M. A study of relative permeability for transient two-phase flow in a low permeability fractal porous medium. *Adv. Geo-Energy Res.* **2018**, *2*, 369–379. [CrossRef]
11. Qin, X.; Zhou, Y.; Sasmito, A.P. An effective thermal conductivity model for fractal porous media with rough surfaces. *Adv. Geo-Energy Res.* **2019**, *3*, 149–155. [CrossRef]
12. Zeng, J.; Liu, J.; Li, W.; Leong, Y.K.; Elsworth, D.; Guo, J. Evolution of shale permeability under the influence of gas diffusion from the fracture wall into the matrix. *Energy Fuels* **2020**, *34*, 4393–4406. [CrossRef]
13. Li, W.; Liu, J.; Zeng, J.; Leong, Y.K.; Elsworth, D.; Tian, J.; Li, L. A fully coupled multidomain and multiphysics model for evaluation of shale gas extraction. *Fuel* **2020**, *278*, 118214. [CrossRef]
14. Li, W.; Liu, J.; Zeng, J.; Leong, Y.K.; Elsworth, D. A fully coupled multidomain and multiphysics model for shale gas production. In Proceedings of the 5th ISRM Young Scholars' Symposium on Rock Mechanics and International Symposium on Rock Engineering for Innovative Future, Okinawa, Japan, 1–4 December 2019.
15. Liu, G.; Yu, B.; Ye, D.; Gao, F.; Liu, J. Study on evolution of fractal dimension for fractured coal seam under multi field coupling. *Fractals* **2020**, *28*, 2050072. [CrossRef]
16. Zeng, J.; Wang, X.; Guo, J.; Zeng, F.; Zeng, F. Modeling of heterogeneous reservoirs with damaged hydraulic fractures. *J. Hydrol.* **2019**, *574*, 774–793. [CrossRef]
17. Raeini, A.Q.; Bijeljic, B.; Blunt, M.J. Generalized network modeling: Network extraction as a coarse-scale discretization of the void space of porous media. *Phys. Rev. E* **2017**, *96*, 013312. [CrossRef]

18. Shaina, K.; Hesham, E.; Carlos, T.; Balhoff, M.T. Assessing the utility of FIB-SEM images for shale digital rock physics. *Adv. Water Resour.* **2015**, *18*, 302–316.
19. Hajizadeh, A.; Safekordi, A.; Farhadpour, F.A. multiple-point statistics algorithm for 3D pore space reconstruction from 2D images. *Adv. Water Resour.* **2011**, *34*, 1256–1267. [CrossRef]
20. Civan, F.; Rai, C.S.; Sondergeld, C.H. Shale-gas permeability and diffusivity inferred by improved formulation of relevant retention and transport mechanisms. *Transp. Porous Media* **2011**, *86*, 925–944. [CrossRef]
21. Zhang, T.Y.; Li, D.L.; Yang, J.Q. A study of the effect of pore characteristic on permeability with a pore network model. *Pet. Sci. Technol.* **2013**, *31*, 1790–1796. [CrossRef]
22. Ju, Y.; Zheng, J.; Marcelo, E.; Sudak, L.; Wang, J.; Zhao, X. 3D numerical reconstruction of well-connected porous structure of rock using fractal algorithms. *Comput. Meth. Appl. Mech. Eng.* **2014**, *279*, 212–226. [CrossRef]
23. Gong, L.; Nie, L.; Xu, Y. Geometrical and Topological Analysis of Pore Space in Sandstones Based on X-ray Computed Tomography. *Energies* **2020**, *13*, 3774. [CrossRef]
24. Iassonov, P.; Gebrenegus, T.; Tuller, M. Segmentation of X-ray computed tomography images of porous materials: A crucial step for characterization and quantitative analysis of pore structures. *Water Resour. Res.* **2009**, *45*, W09415. [CrossRef]
25. Liu, K.; Ostadhassan, M.; Zhou, J.; Gentzis, T.; Rezaee, R. Nanoscale pore structure characterization of the Bakken shale in the USA. *Fuel* **2017**, *209*, 567–578. [CrossRef]
26. Yao, J.; Wang, C.C.; Yang, Y.F.; Sun, H. The Construction method and microscopic flow simulation of carbonate dual pore network model. *Sci. Sin. Phys. Mech. Astron.* **2013**, *43*, 896–902. [CrossRef]
27. Zhang, Z.; Shi, Y.; Li, H.; Jin, H. Experimental study on the pore structure characteristics of tight sandstone reservoirs in Upper Triassic Ordos Basin China. *Energy Explor. Exploit.* **2016**, *34*, 418–439. [CrossRef]

Publisher's Note: MDPI stays neutral with regard to jurisdictional claims in published maps and institutional affiliations.

Article

A Study on the Structure of Rock Engineering Coatings Based on Complex Network Theory

Dayu Ye [1,2], Guannan Liu [1,2,3,*], Feng Gao [1,2], Xinmin Zhu [1,2] and Yuhao Hu [1,2]

[1] State Key Laboratory for Geomechanics and Deep Underground Engineering, China University of Mining and Technology, Xuzhou 221116, China; theoye@cumt.edu.cn (D.Y.); fgao@cumt.edu.cn (F.G.); TS18030215P31@cumt.edu.cn (X.Z.); TS20030008A31@cumt.edu.cn (Y.H.)

[2] Mechanics and Civil Engineering Institute, China University of Mining and Technology, Xuzhou 221116, China

[3] Laboratory of Mine Cooling and Coal-heat Integrated Exploitation, China University of Mining and Technology, Xuzhou 221116, China

* Correspondence: guannanliu@cumt.edu.cn

Received: 15 October 2020; Accepted: 23 November 2020; Published: 26 November 2020

Abstract: As one of the most basic materials of engineering coatings, rock has complex structural characteristics in its medium space. However, it is still difficult to quantitatively characterize the microstructure of rock coatings, such as connectivity and aggregation degree. For this paper, based on a CT-scan model of rock coating, we extracted the network topology of a rock coating sample and verified that its microstructure parameter distribution accords with Barabasi and Albert (BA) scale-free theory. Based on this result, relying on the BA scale-free theory, a dual-porosity network model of rock coating was constructed. We extracted the network structure of the model to verify it, and analyzed the distribution of the microstructure parameters of the model, such as degree distribution, average path length and throat length distribution. At the same time, we analyzed the evolution trend of the permeability of the coating model with the microscopic parameters. Then we discuss the influence of the change of structural parameters on the microstructure of the coating model, and compared with the mainstream rock models at the present stage, the rationality and accuracy of the models are analyzed. This provides a new method for studying the mechanical and permeability properties of engineering coating materials.

Keywords: structure of engineering coatings; pore network model; complex network theory; geometric distribution

---

1. Introduction

Engineering coating refers to the deposition or combination of functional materials on the surface of basic solid materials. The properties and effectiveness of engineering coatings are higher than that of basic solid materials. As coatings, rock materials are widely used in engineering. For example, in order to prevent the deformation or collapse of surroundings, the supporting lining is built with rock and other materials around the tunnel body. In many cases, the structural characteristics are the key factors affecting the mechanical and physicochemical properties of coatings [1,2]. Therefore, it is of great significance to analyze the structural characteristics of natural rocks to improve the mechanical properties of rock coatings [3]. As the mainstream research method of rock microstructure, the pore network model has been widely used. The structure of rock coating can be regarded as a pore network model, and the microstructure of rock is characterized by a pore structure connected by throats. However, rock formation is dense and the pore shape is complex, so it is still difficult to accurately characterize the rock structure and describe the pore connection mode in the pore network models [4]. In recent decades, a great deal of research has been carried out on the microstructure and physical properties of rock mass [5–9], and different kinds of rock numerical models have been established. However, due to the complexity of pore distribution in macrostructure (i.e., geometric structure of network) and micro-structure (i.e., the connection between single pore and other pores), it is still challenging to accurately describe the microstructure of rock coatings.

A key difficulty is that the throats connecting pores are often irregular and unevenly distributed, and the structural characteristics are not uniformly distributed. At present, the characterization of the rock structure mainly adopts the Bieuler number, or macroaverage perspective, which is based on the regular network. Experimental results show that rock pore structure is highly complex, and regular network models and current mainstream real-rock topology models still have great difficulties in characterizing pore connectivity and other topological structures [10–12].

With the advent of the small world model [13] and scale-free model [14], the theoretical framework of modern complex networks was initially formed. In the past 10 years, as a new interdisciplinary field, complex network theory has penetrated into many disciplines from life sciences to physics [15,16]. In previous studies, we found that the porosity degree distribution in the pore structure network of rocks shows the trend of power-law distribution, where the degree of a pore is the number of the pore connected with other pores. That is, with the increase of the number of connected other pores, the number of the pore tends to decrease in power-law ratio. And in the pore network structure of rock, those pores which connect with more other pores play a more important role [17]. Therefore, the scale-free network, that is, the connection network with power-law characteristics of pore degree distribution, can reasonably describe the topological structure of a rock coating network.

The following are several pore network research models that are used to characterize the geometric structure and microscopic characteristics of rocks. The purpose of the model and the research results are summarized in Table 1.

Table 1. Research models used to characterize the microscopic characteristics of rocks.

Author	Years	Purpose of Model	Research Result
Blunt et al. [11].	2002	Research on microscopic effective transport properties based on coordination number rules	A network model that uses additional coordination numbers and is simulated as having irregular connectivity is established.
Hou et al. [18].	2006	Establish a microstructure model considering pore throat radius and shape factor	Figured out that the pore throat radius and shape factor were the main factors affecting polymer retention.
Wang et al. [19].	2007	Explore the influence of reservoir pore structure parameters on water saturation	A network model which can reflect the pore structure of a reservoir is established.
Yao and Tao [20]	2007	Build 3D cubic network model	A mathematical model for solving capillary pressure and relative permeability is established.
Ghous [21]	2008	Explore the influence of pore structure characteristics on rock resistivity	The effects of the structural characteristics of macropores and micropores on rock resistivity at different scales are obtained.
Wang et al. [22].	2013	Establish a multiscale pore network model	Based on the three-dimensional regular network model, a carbonate multiscale network model is constructed.

Based on the porous media [23] and the rock mass digital model [24,25], the Barabasi and Albert (BA) scale-free connection method is used to establish a dual-porosity network model of rock. The porosity degree distribution, average path length, throat length distribution and other parameters of the rock model are used to quantitatively describe the connection of rock pores and throats. The evolution trend of the permeability of the coating model with the microscopic parameters was analyzed. This provides a theoretical basis for the analysis of mechanical properties and physicochemical properties of rock, offering technical support for the more reasonable development and utilization of rock coatings.

2. Characteristics of Rock Pore Network Structure

In this section, we first introduce the research method of rock coating pore network and the mainstream pore network model at the present stage. We made four kinds of rock samples with different porosity and analyzed the microscopic parameters [17]. It is verified that the microstructure of the coated rock accords with the BA scale-free connection mode, as shown in Figure 1.

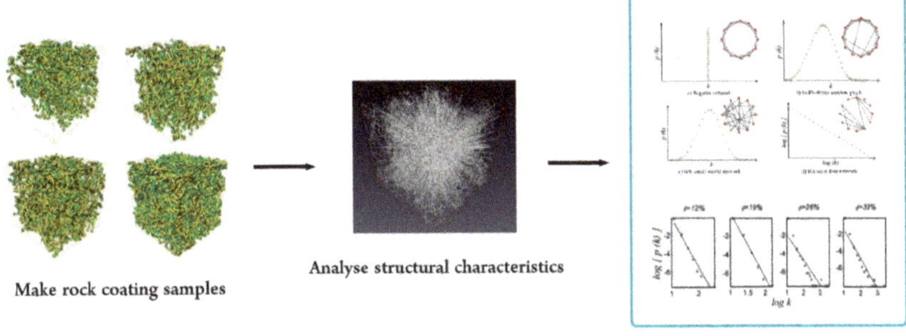

Figure 1. Verification of pore structure of rock coating.

In recent years, as an emerging interdisciplinary field, complex network theory has gradually penetrated into biology, sociology and many other disciplines [26]. Regular network, ER (Erdős–Rényi) random network and WS (Watts-Strogatz) small world network are three classic models describing the structure of random networks. The structure of rock can be regarded as a pore network model, and its microstructure is the pores connected by the throats. As shown in Figure 2.

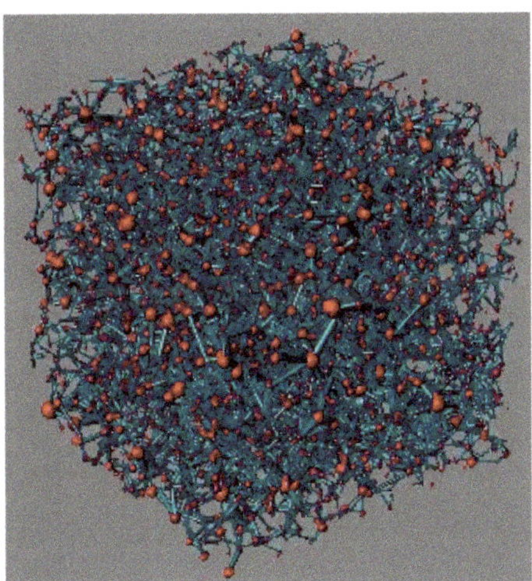

Figure 2. Pore network structure of rock coating.

Regular network, as shown in Figure 3a, the network degree k (the number of edges connected to the node) is the same value, and the regular network has the smallest average path length in all networks with the same number of nodes (the average of all nodes short distance) $L = 1$, $P(k)$ represents the probability of the node. In ER random network, as shown in Figure 3b, the connection probability of two nodes is P, regardless of whether they have a common neighbor node. The degree of most nodes is k, the degree distribution conforms to the Poisson distribution, and the average path length is relatively high.

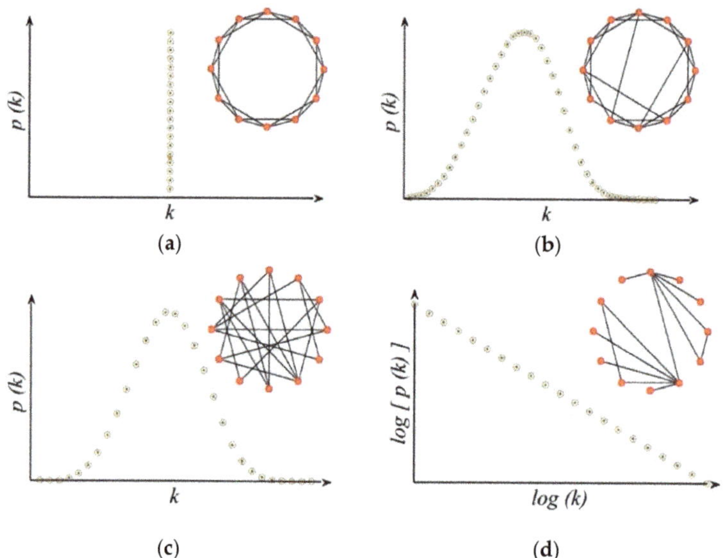

Figure 3. Different network model structures. (**a**) Regular network; (**b**) Erdős–Rényi random graph; (**c**) Watts-Strogatz small world network; (**d**) Barabasi-Albert scale-free network.

As a transition from a regular network to a random network, Figure 3c shows the degree distribution of the WS small world network. Most nodes are not connected, but different nodes can be reached by fewer edges.

Empirical research shows that other real networks such as the mobile internet and cellular metabolic networks are irregular networks, and the connectivity distribution function takes the form of a power rate [27]. In order to explain the generation mechanism of the power-law distribution, Barabasi and Albert proposed a scale-free network model, namely the BA scale-free model [14]. As a model for describing heterogeneous networks, the degree distribution function of the BA scale-free network is:

$$P(k) = \frac{2m(m+1)}{k(k+1)(k+2)} \propto 2m^2 k^{-3} \tag{1}$$

where m is the number of new nodes connected to existing nodes during network expansion, namely:

$$P(k) \approx k^{-\gamma} \tag{2}$$

Among them, γ is power exponent, and there is no obvious characteristic value (Figure 3d), but it is extremely robust to random node failures. As long as the few nodes with the largest degree in the network are consciously removed, it will have a huge impact on the connectivity of the entire network.

In the previous research [17], we used the pressing method to produce artificial sandstone with porosities of 12%, 19%, 26% and 33% through different ratios of 50–300 mesh natural sandstone powder and epoxy resin. An X-ray synchrotron radiation imaging experimental device with an effective pixel resolution of 12–14 μm is used to scan and generate a tomographic image stack. Then, the pore network model is established by using the ossification algorithm to take the pore as the node and the throat as the edge to extract the pore structure information. As shown in Figure 4. At the same time, in order to study the microstructure of different structural rocks, we analyze the characteristics of the rock network with different porosity.

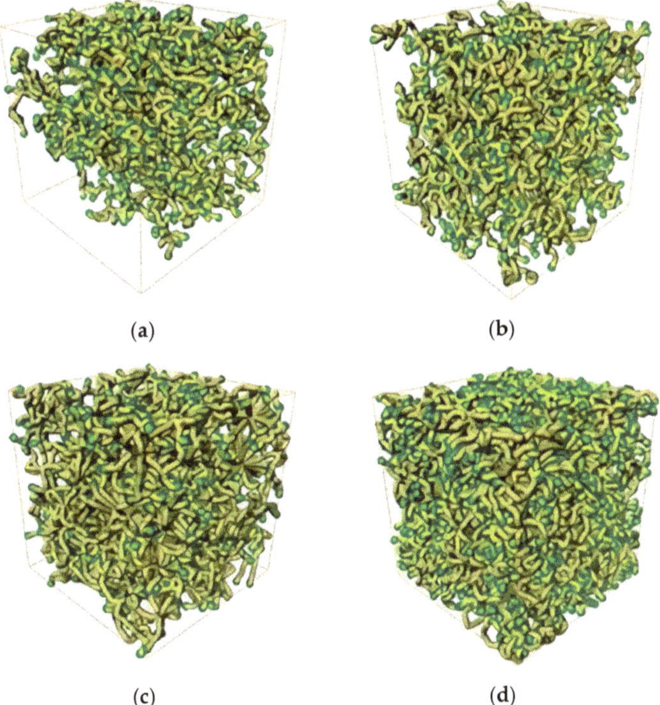

Figure 4. Three-dimensional sandstone pore network. (**a**) porosity = 12%; (**b**) porosity = 19%; (**c**) porosity = 26%; (**d**) porosity = 33%.

The distribution function $P(k)$ is used to represent the distribution of network node degrees. $P(k)$ represents the probability that the randomly selected node degree is k. In the rock coating network model, the degree of a pore is the number of the pore connected with other pores. The experimental results show that the pore network of different degrees satisfies the power-law distribution, as shown in Figure 5. Therefore, the rock network conforms to the BA scale-free network, that is, there are a small number of scale-free network height nodes called the "hub" of the network. In a sense, these joints may play a leading role in the deformation and failure process of rock under load and stress distribution [28,29].

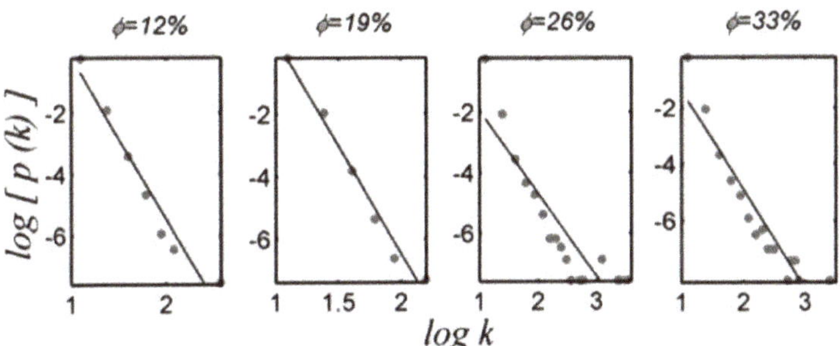

Figure 5. Distribution curve of degree k of sandstone structure.

3. Construction of Three-Dimensional Rock Pore Network Model Based on BA Scale-Free Connection

The construction of the pore network model retains the properties of the real macropore network model and the micropore network model as much as possible, and the geometric topology of the pore network model is an important evaluation criterion to ensure the accuracy of the model. Wang proposed a theoretical method to integrate and construct a dual-porosity network model [30]. First, determine the spatial domain Ω of the large-pore network reconstruction, that is, determine the geometric size of the dual-porosity network model. Based on a real-rock model size, it is assumed that the number of pores is N_p, the network volume is $V = lwh$ and $l = w = h$. The number of macropores in the established pore network is N_p', the network volume is $V' = LWH$ and $L = W = H$. Therefore, the size factor is:

$$\xi = L/l = W/w = H/h \tag{3}$$

Then we can obtain:

$$N_p' = \xi^3 N_p \tag{4}$$

We introduce the BA scale-free connection method in the complex network theory to generate and connect the nodes in the pore domain. BA scale-free network meets the following two characteristics:

- Growth: Starting from a network with m_0 nodes, each time a new node is introduced and connected to m existing nodes, where $m \leq m_0$.
- Optimal connection: The probability $\prod_i$ of a new node j connected to an existing node i, the degree k_i of node i, and the degree k_j of node j satisfy the following relationship:

$$\prod_i = \frac{k_i}{\sum_j k_j} \tag{5}$$

After t steps, the algorithm will generate a network with $N = t + m_0$ nodes and mt edges, that is, the program will generate a pore network model with N pores and mt throats. Changing the parameters of m_0 and m in the BA scale-free network will change the structure of the pore network model accordingly.

At the same time, the average path length of the BA scale-free network structure, that is, the average length of the throats, has the following relationship:

$$L = \frac{1}{\frac{1}{2}N(N+1)} \sum_{i \geq j} d_{ij} \tag{6}$$

where d represents the shortest path length between any two nodes (the minimum number of edges connecting any two nodes), and N is the total number of network pores.

During the process from m_0 to N_p', the generated pore cannot coincide or overlap with the existing pore. That is, when a new pore k with a radius of r_k is generated, it should satisfy that it will not cover the generated pore $j(j = 1, 2, ..., k-1)$, that is:

$$r_k + r_j < d_{jk} \tag{7}$$

where d_{jk} is the Euclidean distance between pores k and j. It is only necessary to determine the radius of the new pore and the pore radius which has been generated in the adjacent small area around it.

According to the distribution of the pore radius of the real-rock model, we determined the value range of the large pore radius in the dual-porosity model. And randomly generate large pores with a radius within the value range at the pore center coordinate point of the large pore space domain.

So far, the macropore network model based on the BA scale-free network has been established. The filling of the micropore network refers to the generation of micropores and microthroat unit bodies based on the BA scale-free network connection method in the generated macropore network skeleton space. It mainly includes the steps of determining the spatial domain of the micropore network, generating equivalent micropores, and constructing throats between micropores.

The micropore network cannot overlap with the macropore network during the filling process, so the space domain filled by the micropore network is the skeleton space of the macropore network. Assuming that the overall space of the pore network model is Ω, the pore space of the constructed macropore network is Ω_1, and the space domain of the micropore network model to be constructed is Ω_2. Then the micropore space domain can be expressed as:

$$\Omega_2 = \Omega - \Omega_1 \tag{8}$$

Analogous to Equation (4), we can get the number of micropores to be constructed in the micropore network:

$$N_{pm}' = \xi^3 N_{pm} \tag{9}$$

Among them, N_{pm} is the number of micropores in the real-rock digital model.

Similarly, we use the BA scale-free connection method (Equation (5)) to generate the microporous network model. The connection method of the micro throat is similar to that of the large throat. At the same time, before a new micropore is generated, it must be ensured that it will not cover the micropores that have been generated. The judgment condition is the same as Equation (7).

However, at this time, the newly generated macropore network and the micropore network are not connected, so it is necessary to add a connecting throat between the macropore and the micropore to integrate the dual-porosity network. Wang [30] proposed a connection construction method that integrates a two-scale network model. At the center of the original micropore network, place the pore body with the smallest radius in the macropore network, and calculate the coordination number between the surface of this pore and the surrounding micropore throat. By increasing the radius of the pore body, the coordination number of a series of pore surface and the surrounding micropore throat can be obtained, and then the coordination number distribution map between the macropore and the micropore can be obtained. According to the coordination number distribution map between the macropores and the micropores, the connecting throats between the macropores and the adjacent micropores are added.

Through the above steps, the established macroporous network based on the BA scale-free connection mode can be integrated with the microporous network, and then a dual-porosity network model based on the BA scale-free network connection mode is constructed.

4. Results and Analysis

We use the Carbonatite digital model proposed by the Martin Blunt team of Imperial College London [31]. At the same time, based on the size of the real-rock model, combined with Equations (3), (4), (8) and (9), the scale factor ξ is selected. We determined the number of pores in the macropore network model (N_p') and the number of pores in the micropore network model (N_{pm}'). Combining the BA scale-free connection principle and the above equations, we constructed a dual-porosity rock coating network model. As shown in Figure 6.

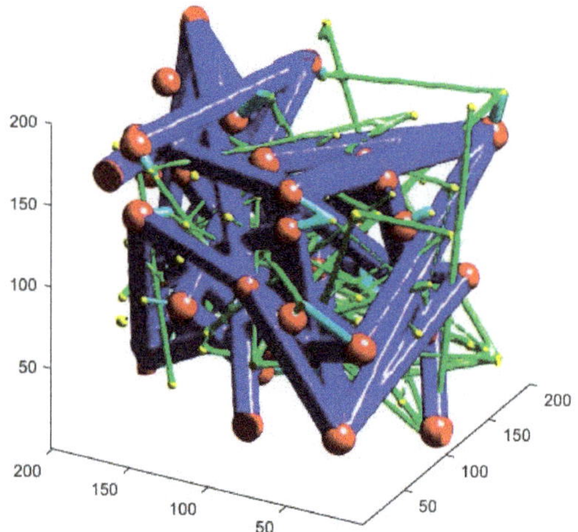

Figure 6. Two-scale network model based on BA scale-free network connection.

To verify the correctness of the model, we extracted the digital structure characteristics of the pore network of the real-rock digital model, and analyzed the degree distribution of the pore network model, as shown in Figure 7.

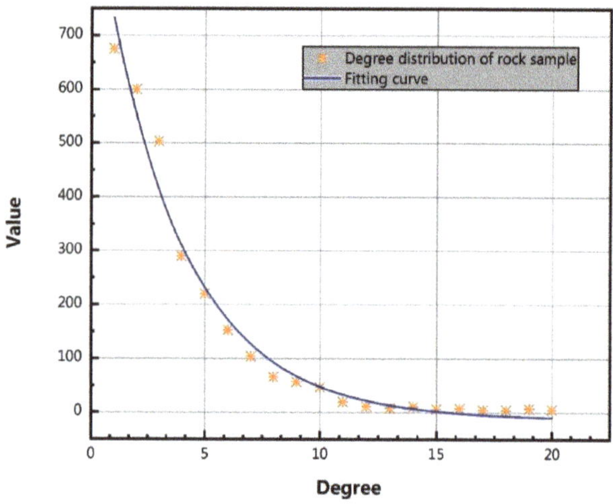

Figure 7. Distribution characteristics of the carbonate sample.

It can be seen from Figure 7 that the distribution characteristic of the real-rock pore network degree accords with the property of power ratio distribution. The characteristic of its distribution function accords with the Equations (1) and (2), that is, the BA scale-free network connection is the main connection mode of the rock pore network.

Similarly, we analyzed the structural characteristics of the dual-porosity rock coating network model based on the BA scale-free network, and analyzed its degree distribution characteristics. Using the size factor method proposed by Wang [30], the model degree distribution value is multiplied by the size factor obtained above, and the revised model degree distribution trend is obtained. The revised model degree distribution characteristics and real-rock degree distribution characteristics are shown in Figure 8.

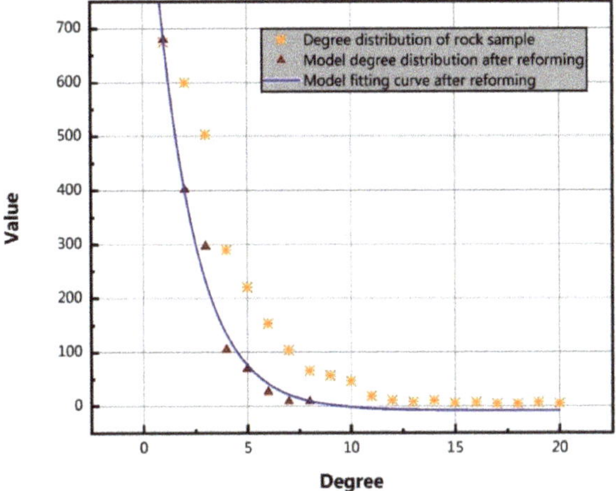

Figure 8. Comparison of the revised model and real-rock degree distribution characteristics.

It can be seen from Figure 8 that the degree distribution characteristics of the dual-porosity network model proposed in this paper are in good agreement with those of the real-rock sample. In order to further verify the correctness of the model proposed in this paper, we selected four different rock samples. By extracting the network structure and parameter analysis of them, we have obtained the degree distribution characteristics. At the same time, by changing the specific parameters m_0 and m in the BA scale-free network (Equation (5)), we established four different dual-porosity network models and analyzed the structural characteristics of the network models. The degree distribution characteristics of the four different network models and the rock samples are shown in Figure 9.

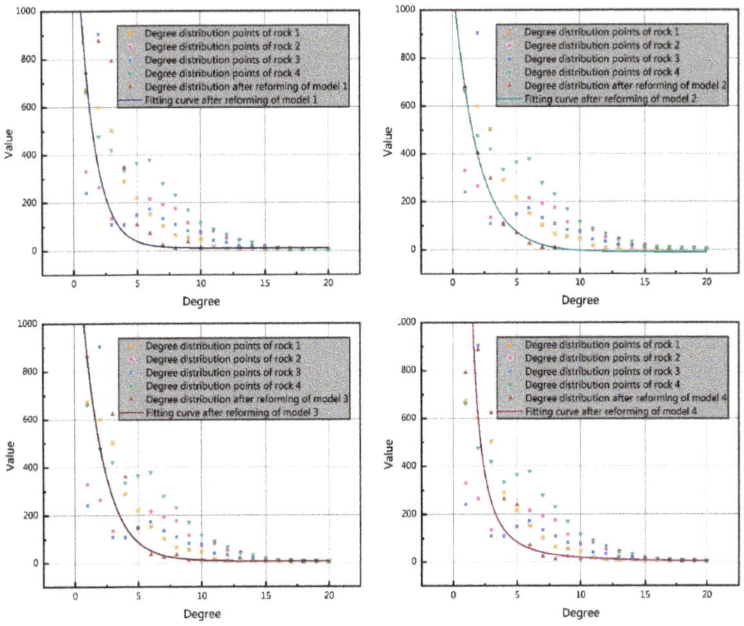

Figure 9. Comparison of distribution of four different rocks with different models.

From Figure 9, we can see that although the fitting curve of the model changes with the change of m_0 and m, it is still in good agreement with the degree distribution characteristics of the rock samples. That is, the model proposed in this paper can well describe the structural characteristics of the rock coating pore network.

Meanwhile, we explored the effects of different degree distributions on network permeability. We analyzed the degree distribution of four different rocks and calculated the network permeability of these four rocks, as shown in Figures 10 and 11.

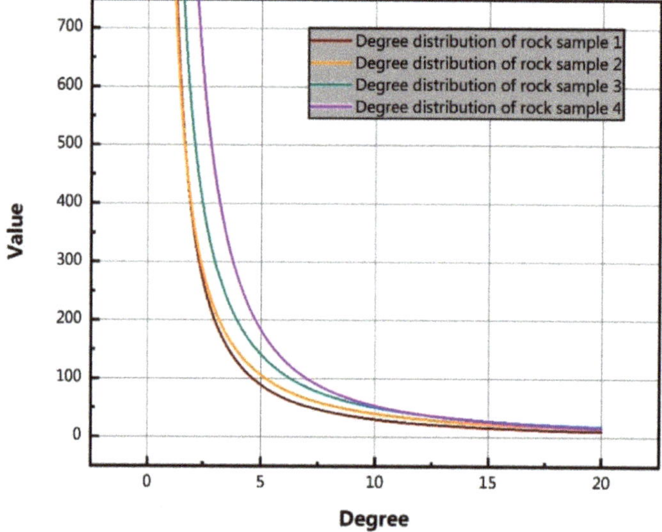

Figure 10. Degree distribution of four different rocks.

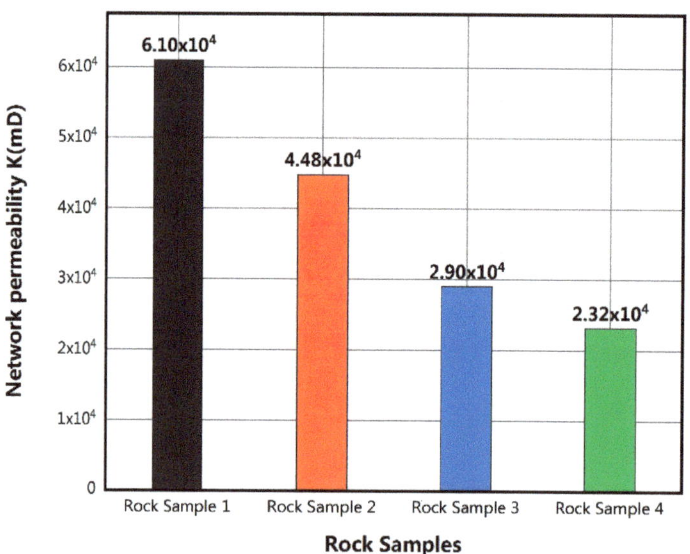

Figure 11. Network permeability of different samples.

From Figures 10 and 11 we can see that the permeability of the rock network model is closely related to its degree distribution. With the increase of the number of pores with small degree in the rock network model, the overall permeability decreases. If the degree of a pore is small, the number of throats connected by this pore is small, and the number of pores connecting other pores is also small. In the process of rock seepage, with the increase of the number of pores with weak connectivity, the seepage intensity decreases and the permeability also decreases.

Then, we analyzed the influence of m_0 and m on the rock structure. First of all, we analyzed the influence of structural parameters on the degree distribution of the network model. Since macropores play a leading role in the process of load and stress distribution of rock, in order to simplify the calculation, we constructed a two-dimensional large pore network model equivalent to the real-rock size. After that, we changed the value of m_0 and m, analyzed the degree distribution of each model, and compared it with the degree distribution of rock samples. The degree distributions of the models under different m_0, m and the degree distribution of the rock sample are shown in Figure 12.

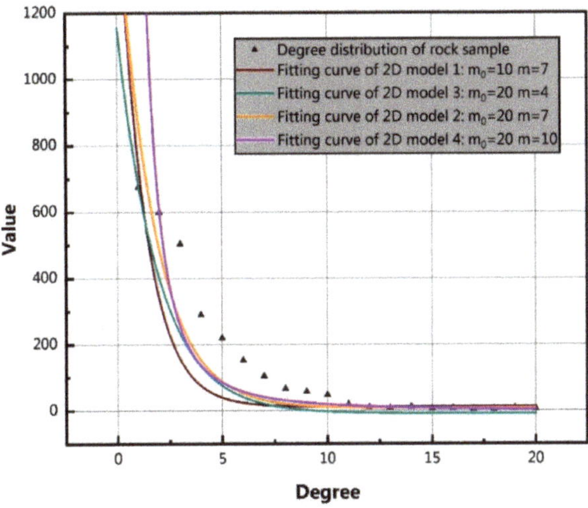

Figure 12. The characteristics of degree distributions under different m_0 and m.

From Figure 12, it can be seen that the degree distributions under different m_0 and m strictly obey the power-law distribution, and are in good agreement with the distribution characteristics of the rock sample degree. When m and other microstructure parameters remain unchanged, the number of the pores with the degree between two and seven increases with the increase of m_0. When other parameters remain unchanged, with the increase of m_0, the number of pores with degrees one and two increases sharply. And the number of pores with a higher degree value does not change much with the change of m_0 and m.

The average path length of the rock coating model, that is, the average length of the throat connecting the rock pores in the pore network, plays an important role in analyzing the stress distribution and failure characteristics of the model. When other parameters such as rock porosity, size and structure remain unchanged, based on Equation (6). We analyzed the evolution of the average path length of the model under different values of m_0 and m, as shown in Figure 13.

From Figure 13, we can conclude that the average path length of the model increases with the increase of m_0 and m, when other parameters are identical. And m has a greater impact on the average path length. With the increase of the average path length, there are more throats in the network. This will lead to the uneven stress distribution of rock under loading, which is more likely to cause the failure of rock coating. Therefore, we speculate that from the perspective of the average path length, the increase of m_0 and m is not conducive to the stress process of the rock coating.

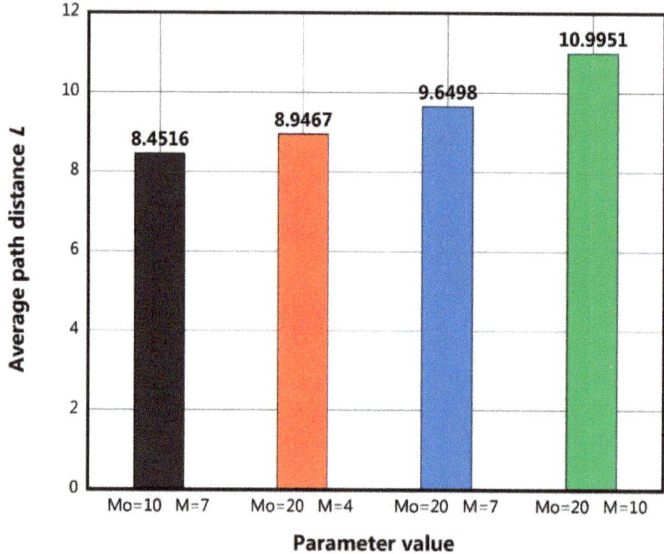

Figure 13. Evolution of the average path length under different values of m_0 and m.

In order to further explore the influence of the value of m_0 and m on the coating, we explored the distribution of rock throat length. We selected the above-mentioned Carbonatite digital model proposed by Martin Blunt's team [31] based on CT tomography, extracted the network structure and analyzed its throat distribution, as shown in Figure 14.

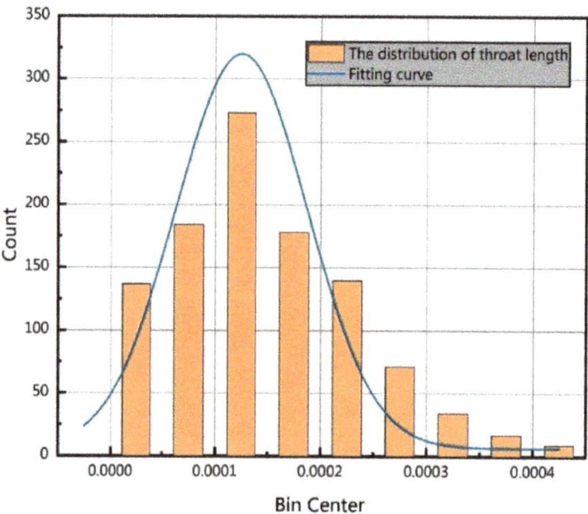

Figure 14. The throat length distribution characteristics of carbonate samples.

From Figure 14, we can conclude that the distribution of rock throat length roughly presents a Gaussian distribution. The length of the throats is mostly concentrated in the range of (0.00005–0.00015), and the number of throats with larger length is much smaller than that of those with smaller length.

Similarly, we changed the values of m_0 and m, and analyzed the throat length distribution of each model, as shown in Figure 15. The network parameters such as the size and the number of pores of each model are the same as those of the rock samples.

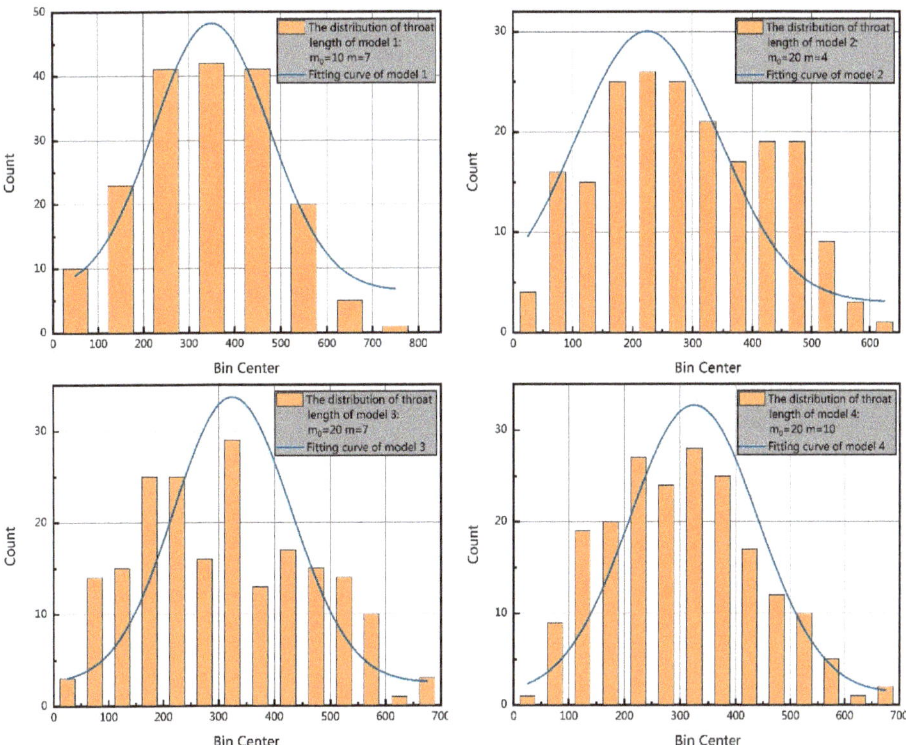

Figure 15. The distribution characteristics of throat length under different values of m_0 and m.

Figure 15 shows the effect of m_0 and m on the throat length distribution of the rock coating. The throat length distribution characteristics of the four models are consistent with the distribution of the rock sample. When the value of m_0 is 10, the throat length is mainly concentrated in the range of 250–550, and there are throats of larger length (greater than 700). With the increase of m_0, the length of the throats is mainly distributed in the range of 75–450. That is, as m_0 increases, the throat length generally shows a decreasing trend. At $m = 4$, the length of throats is small (75–350), as m increases, the concentrated interval of throats length also increases. That is, with the increase of m, the overall throat length shows an increasing distribution trend.

Then, we analyzed the degree distribution of different rock network methods in the current mainstream models. The ER random network, WS small world network and conventional regular rock connection models were selected.

Where the degree distribution of the ER random graph satisfies [32]:

$$P(k) = \binom{N-1}{K} P^k (1-P)^{N-1-k} \approx \frac{\langle k \rangle^k}{k!} e^{-\langle k \rangle} \tag{10}$$

where $P(k)$ is the probability that the degree of a pore is k, and N is the total number of pores in the network. The construction algorithm of WS small world model is as follows [13]:

(1) Start modeling with a ring of regular networks: The network contains N nodes, and each node connects to the K nodes closest to it. Where $N >> K >> ln(N) >> 1$.

(2) Randomize reconnection: reconnect each edge in the network randomly with probability p, that is, one endpoint of the edge remains unchanged, while the other endpoint is taken as a randomly selected node in the network. It stipulates that there can be at most one edge between any two different nodes, and each node cannot have an edge connected to itself.

And the total number of nodes in different networks is consistent with the number of pores in the rock samples. We compared the degree distribution of different models with that of the rock sample, as shown in Figure 16.

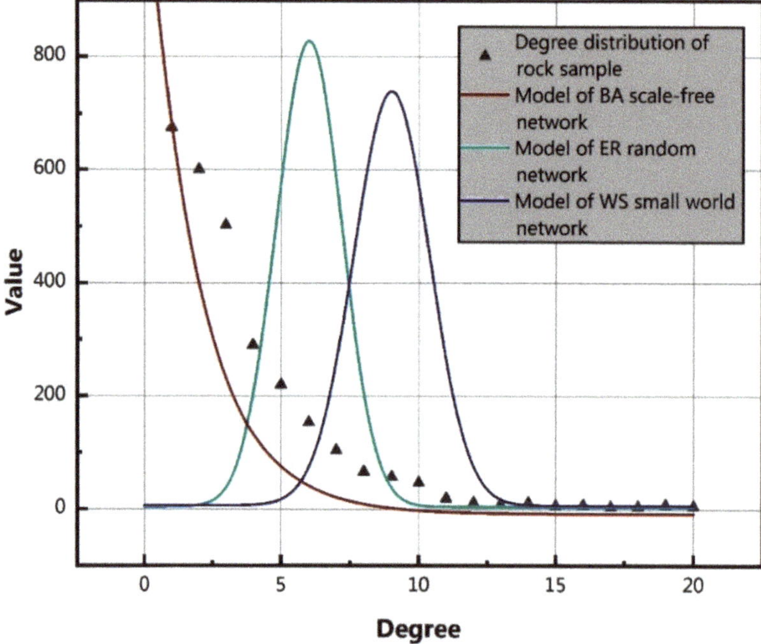

Figure 16. Comparison of different models with real-rock distribution characteristics (note: the degree distribution of regular model is the points of degree = 2, 3, 4, and 6).

From Figure 16, we can conclude that the current conventional pore network model is not in good agreement with the structural characteristics of the rock sample compared with the BA scale-free connection model, where the degree distribution characteristics of the real-rock sample is power-law distribution, and the rock coating model proposed in this paper is also power-law distribution. However, the degree distribution characteristics of the ER random network and WS small world network models are Gaussian distribution, and the regular rock connection model is point distribution, which do not match the degree distribution characteristics of the real-rock sample.

Therefore, the pore network model based on a BA scale-free connection can describe rock structure more accurately. It provides theoretical basis and technical support for the analysis of rock mechanical properties and physicochemical properties, and more reasonable development and utilization of rock coatings. At the same time, because we study the microstructure of rock coatings, the analysis of the entire coating requires a large number of samples, which may have cost-effective limitations. Although the model is not compared with the permeability results in the published study, the microstructure has been compared with the actual rock samples, which is Figures 8 and 9.

In addition, this study uses the method of compiling the numerical model of a pore network to simulate the actual coated rock, so there is no error analysis, which may cause numerical limitations. Since we only consider the characteristics of rock microstructure, in the next study, we will explore the influence of microparameters (such as degree distribution, average path length, throat length, etc.) on the mechanical and physicochemical properties of the overall coating.

5. Conclusions

The rock pore structure is an important factor of coating failure. In order to analyze the microstructure of rock coating more accurately, a dual-porosity rock structure model was established in this paper. The pore connection structure was constructed by a BA scale-free method, and the rock coating model was compared with different types of real-rock samples to verify the correctness. We discussed the influence of structural parameters on the structure and permeability of the network model. The main conclusions are as follows:

- BA scale-free method can accurately and truly describe the pore structure of coated rocks. The dual-pore network model proposed in this paper can well match the structural distribution characteristics of different types of rocks, so as to reasonably describe the structure of the coating.
- With the increase of the number of pores with a small degree in a rock network model, the overall permeability decreases. And the number of the pores with the degree between two and seven increases with the increase of m_0. With the increase of m_0, the number of pores with degree one and two increases sharply, while the number of pores with higher degree value does not change much with the change of m_0 and m. Besides, with the increase of m_0 and m, the average path length of the rock shows an increasing trend. And with the increase of m_0, the overall throat length shows a decreasing trend, while the effect of m is opposite.
- Compared with the mainstream pore network model, the BA scale-free connection method can describe rock structure more accurately. This provides a new method for studying mechanical and permeability properties of engineering coating materials.

Author Contributions: Methodology, software, formal analysis, D.Y.; Conceptualization, resources, funding acquisition, G.L.; Supervision, investigation, F.G.; Data curation, formal analysis, X.Z.; Writing—Review and editing Y.H. All authors have read and agreed to the published version of the manuscript.

Funding: This research was funded by the open fund of The Key Laboratory of Coal-based CO_2 Capture and Geological Storage in Jiangsu (Grant No. 2017B06), the Priority Academic Program Development of Jiangsu Higher education Institutions, china scholarship council and the National Natural Science Foundation of China (11202228).

Conflicts of Interest: The authors declare no conflict of interest.

References

1. Zhang, D.; Huang, Z.; Yin, Z.; Ran, L.; Huang, H. Predicting the grouting effect on leakage-induced tunnels and ground response in saturated soils. *Tunn. Undergr. Space Technol.* **2017**, *65*, 76–90. [CrossRef]
2. Ninić, J.; Meschke, G. Simulation based evaluation of time-variant loadings acting on tunnel linings during mechanized tunnel construction. *Eng. Struct.* **2017**, *135*, 21–40. [CrossRef]
3. Silin, D.; Kneafsey, T.J. Shale Gas: Nanometer-Scale Observations and Well Modelling. *J. Can. Pet. Technol.* **2012**, *51*, 464–475. [CrossRef]
4. Loucks, R.G.; Reed, R.M.; Ruppel, S.C.; Jarvie, D.M. Morphology, Genesis, and Distribution of Nanometer-Scale Pores in Siliceous Mudstones of the Mississippian Barnett Shale. *J. Sediment. Res.* **2009**, *79*, 848–861. [CrossRef]
5. National Research Council (US); Committee on Fracture Characterization, Fluid Flow. *Rock Fractures and Fluid Flow: Contemporary Understanding and Applications*; National Academies Press: Washington, DC, USA, 1996.
6. Baghbanan, A.; Jing, L. Hydraulic properties of fracture rock masses with correlated fracture length and aperture. *Int. J. Rock Mech. Min. Sci.* **2007**, *44*, 704–719. [CrossRef]

7. Koyama, T.; Neretnieks, I.; Jing, L. A numerical study on differences in using Navier-Stokes and Reynolds equations for modeling the fluid flow and particle transport in single rock fractures with shear. *Int. J. Rock Mech. Min. Sci.* **2008**, *45*, 1082–1101. [CrossRef]

8. Zhao, Z.; Jing, L.; Neretnieks, I.; Moreno, L. Numerical modeling of stress effects on solute transport in fractured rocks. *Comput. Geotech.* **2011**, *38*, 113–126. [CrossRef]

9. Liu, R.; Jiang, Y.; Li, B.; Wang, X.; Xu, B. Numerical calculation of directivity of equivalent permeability of fractured rock masses network. *Rock Soil Mech.* **2014**, *35*, 2394–2400.

10. Dillard, L.A.; Blunt, M.J. Development of a pore network simulation model to study nonaqueous phase liquid dissolution. *Water Resour. Res.* **2000**, *36*, 439–454. [CrossRef]

11. Blunt, M.J.; Jackson, M.D.; Piri, M.; Valvatne, P.H. Detailed physics, predictive capabilities and macroscopic consequences for pore-network models of multiphase flow. *Adv. Water Resour.* **2002**, *25*, 1069–1089. [CrossRef]

12. Zhao, H.Q.; Macdonald, I.F.; Kwiecien, M.J. Multi-Orientation Scanning: A Necessity in the Identification of Pore Necks in Porous Media by 3-D Computer Reconstruction from Serial Section Data. *J. Colloid Interface Sci.* **1994**, *162*, 390–401. [CrossRef]

13. Watts, D.J.; Strogatz, S.H. Collective dynamics of 'small-world' networks. *Nature* **1998**, *393*, 440–442. [CrossRef] [PubMed]

14. Barabasiand, A.-L.; Albert, R. Emergence of Scaling in Random Networks. *Science* **1999**, *286*, 509–512. [CrossRef] [PubMed]

15. Sanyal, A.; Lajoie, B.R.; Jain, G.; Dekker, J. The long-range interaction landscape of gene promoters. *Nature* **2012**, *389*, 109–113. [CrossRef] [PubMed]

16. Jiang, Z.; Liang, M.; Li, Q.; Guo, D. Optimal dynamic bandwidth allocation for complex networks. *Phys. A Stat. Mech. Appl.* **2013**, *392*, 1256–1262. [CrossRef]

17. Feng, G.; Guannan, L.; Dayu, Y.; Xin, L.; Xiaoqian, Z. Study on microstructure mechanism of sandstone based on complex network theory. *J. Meas. Eng.* **2020**, *8*, 27–33. [CrossRef]

18. Jian, H.; Zhenquan, L.; Yueming, C. Research on retained polymer distribution laws by three-dimension network model. *Chin. J. Comput. Mech.* **2006**, *23*, 453–458.

19. Wang, K.; Sun, J.; Guan, J.; Su, Y. Percolation network modeling of electrical properties of reservoir rock. *Appl. Geophys.* **2005**, *2*, 223–229. [CrossRef]

20. Jun, Y.; Jun, T.; Aifen, L. Research on oil-water two-phase flow using 3D random network model. *Acta Petrol. Sin.* **2007**, *28*, 94–97.

21. Ghous, A.; Knackstedt, M.A.; Arns, C.H.; Sheppard, A.; Kumar, M.; Sok, R.; Senden, T.; Latham, S.; Jones, A.C.; Averdunk, H.; et al. 3D imaging of reservoir core at multiple scales: Correlations to petrophysical properties and pore scale fluid distributions. In Proceedings of the International Petroleum Technology Conference, Kuala Lumpur, Malaysia, 3–5 December 2008; pp. 3–5.

22. Wang, C.C.; Yao, J.; Yang, Y.F.; Wang, X.; Ji, G.S.; Gao, Y. The construction of carbonate multiscale network model based on regular network. *Chin. J. Comput. Mech.* **2013**, *30*, 231–235.

23. Yu, B.; Cheng, P. A fractal permeability model for bi-dispersed porous media. *Int. J. Heat Mass Transf.* **2002**, *45*, 2983–2993. [CrossRef]

24. Yeong, C.L.Y.; Torquato, S. Reconstructing random media. *Phys. Rev. E* **1998**, *57*, 495–506. [CrossRef]

25. Wu, K.; Van Dijke, M.I.J.; Couples, G.D.; Jiang, Z.; Ma, J.; Sorbie, K.S.; Crawford, J.; Young, I.; Zhang, X. 3D Stochastic Modelling of Heterogeneous Porous Media – Applications to Reservoir Rocks. *Transp. Porous Media* **2006**, *65*, 443–467. [CrossRef]

26. Williams, R. Methodology or chance the degree distributions of bipartite ecological networks. *PLoS ONE* **2011**, *6*, e17645. [CrossRef]

27. Takaguchi, T.; Sato, N.; Yano, K.; Masuda, N. Importance of individual events in temporal networks. *New J. Phys.* **2012**, *14*, 093003. [CrossRef]

28. Kitsak, M.; Gallos, L.K.; Havlin, S.; Liljeros, F.; Muchnik, L.; Stanley, H.E.; Makse, H.A. Identification of Influential Spreaders in Complex Networks. *Nat. Phys.* **2010**, *6*, 888. [CrossRef]

29. Wang, W.; Tang, M.; Zhang, H.F.; Gao, H.; Do, Y.; Liu, Z.H. Epidemic spreading on complex networks with general degree and weight distributions. *Phys. Rev. E* **2014**, *90*, 042803. [CrossRef]

30. Chenchen, W. Construction Theory and Method of Dual Pore Network Model in Carbonate Media. Ph.D. Thesis, China University of Petroleum, Qingdao, China, 2013. (In Chinese).

31. Micro-CT Images and Networks. Available online: https://www.imperial.ac.uk/earth-science/research/research-groups/perm/research/pore-scale-modelling/micro-ct-images-and-networks/ (accessed on 28 October 2020).
32. Bollobas, B. *Random Graphs*, 2nd ed.; New York Academic Press: Washington, DC, USA, 2001.

Article

Effect of Interface Coating on High Temperature Mechanical Properties of SiC–SiC Composite Using Domestic Hi–Nicalon Type SiC Fibers

Enze Jin [1], Wenting Sun [1], Hongrui Liu [1,2], Kun Wu [1], Denghao Ma [1], Xin Sun [1], Zhihai Feng [1], Junping Li [1,*] and Zeshuai Yuan [1]

[1] Key Laboratory of Advanced Functional Composites Technology, Aerospace Research Institute of Materials & Processing Technology, Beijing 100076, China; jinenzelc@163.com (E.J.); wentingsun2008@163.com (W.S.); liu_hongrui@126.com (H.L.); wukunjez@163.com (K.W.); madenghao1987@163.com (D.M.); sunxinhit@163.com (X.S.); fengzhihaijez@163.com (Z.F.); fnl37051123@163.com (Z.Y.)

[2] Science and Technology on Advanced Ceramic Fibers and Composites Laboratory, College of Aerospace Science and Engineering, National University of Defense Technology, Changsha 410073, China

* Correspondence: jpli@iccas.ac.cn

Received: 10 April 2020; Accepted: 11 May 2020; Published: 15 May 2020

Abstract: Here we show that when the temperature exceeded 1200 °C, the tensile strength drops sharply with change of fracture mode from fiber pull-out to fiber-break. Theoretical analysis indicates that the reduction of tensile strength and change of fracture mode is due to the variation of residual radial stress on the fiber–matrix interface coating. When the temperature exceeds the preparation temperature of the composites, the residual radial stress on the fiber–matrix interface coating changes from tensile to compressive, leading to the increase of the interface strength with increasing temperature. The fracture behavior of SiC–SiC composites changes from ductile to brittle when the strength of fiber–matrix interface coating exceeds the critical value. Theoretical analysis predicts that the high temperature tensile strength can increase with a decrease in fiber–matrix interface thickness, which is verified by experiments.

Keywords: ceramic matrix composite; high temperature strength; SiC fiber; fiber–matrix interface coatings; residual stress

---

1. Introduction

Continuous SiC fiber-reinforced SiC matrix composite (SiC–SiC) is known as a high-temperature resistant material, with excellent properties such as high specific strength, high specific stiffness, high temperature resistance, long-term oxidation resistance and erosion resistance. Therefore, SiC–SiC composite has a wide application prospect in the thermal protection system of aerospace vehicles and hot-end components of aircraft engines [1–3]. In addition, SiC–SiC composite with near stoichiometric (3rd generation) SiC fibers has high irradiation resistance, which is one of the most promising materials for structural application in nuclear fusion reactor and nuclear fission power system [4–6].

In the view of fracture mechanics of composites, fiber–matrix interface is the most important factor that determines the structural stability of a composite. Recently, intensive efforts have been devoted to study the influences of interface on mechanical properties of SiC–SiC composites. Wang et al. studied the tensile creep properties of 2D-SiC–SiC composites in vacuum at high temperature [7]. The results show that the interface strength is relatively strong and the matrix cracking is inhibited at 1300 and 1350 °C. The roles of fiber on the creep behaviors of composites are determined by the fiber–matrix interface. Buet et al. systematically analyzed the fiber–matrix interface of SiC–SiC reinforced by stoichiometric SiC fibers. Their experiments demonstrate that the interface shear stress depends on the texture of the carbon interface. Highly anisotropic pyrocarbon interfaces increase the interfacial shear stress [8]. They also found that Tyranno SA3 fiber has a granular and rough surface which leads to an increase of the interface shear strength. The Tyranno SA3-based composites with stronger interface exhibit a brittle behavior [9]. Fellah et al. investigated the influence of the carbon interface on the mechanical behavior of SiC–SiC [10]. The results show that the fiber–matrix debonding behavior depends strongly on the nature of the carbon on the SiC fiber surface, which is different according to the SiC fiber. Hsu et al. measured the interface tensile strength of SiC/Si directly for the first time by taking advantage of the FIB method [11].

It is well acknowledged that appropriate interface can effectively transfer load and inhibit crack propagation. Ding et al. studied the influence of PIP–SiC interface on the mechanical properties of SiC–SiC composites at high temperature [12]. The results show that the PIP–SiC interface can improve the toughness of the composite by deflecting most cracks in matrix. Wang et al. developed SiC–SiC composites with a PyC–SiC multilayer interface [13]. The results reveal that when the composites are oxidized at the temperature higher than 900 °C, the composites exhibit self-healing characteristics. Shimoda et al. developed non-brittle fractures in SiC–SiC composites without a fiber–matrix interface [14]. They found that sandwiched layers with a porous matrix in a laminate can achieve crack deflection and exhibit ductile fracture behavior. Therefore, in order to provide a deep insight on the stability of SiC–SiC composite, it is important to investigate the effect of interface coating on the mechanical properties.

Domestic Hi–Nicalon type SiC fiber is a new material and there are few studies on the its composite. In this study, we investigate the tensile properties of SiC–SiC composites at high temperature reinforced by domestic Hi–Nicalon type SiC fiber and illuminate effect of interface coating on the mechanical property variations by combining theoretical analysis and experiments.

2. Materials and Methods

2.1. Preparation of SiC–SiC Composites

Two-dimensional preform was woven using domestic Hi–Nicalon type SiC fibers. The SiC fibers were provided by Xiamen University (Xiamen, China). The fiber diameter is 14 μm, density is 2.79 g/cm, tensile strength is 2.7 GPa, and the modulus is 270 GPa. The typical chemical composition of the fiber is listed in Table 1. The properties of the domestic Hi–Nicalon type SiC fiber are close to those of Hi–Nicalon fiber [15,16]. Pyrolytic carbon (PyC) layers with a thickness of about 580 nm were coated on the surface of the preforms by chemical vapor deposition (CVD) method. SiC–SiC composites were prepared by precursor infiltration and pyrolysis (PIP) process. The preforms were impregnated with liquid state Polycarbosilane (PCS) by a vacuum infiltration method and pyrolyzed at 850 °C in an inert Argon atmosphere. The impregnation and pyrolysis process were repeated 10 times until weight increase was less than 1%. The final porosity of the composites is 6%–9%.

Table 1. Typical chemical composition of domestic Hi–Nicalon type SiC fibers.

Si Content Wt %	C Content Wt %	O Content Wt %	C/Si Mole Ratio
61.5	37.9	0.6	1.44

2.2. Tensile Test of SiC–SiC Composites at High Temperature in Air

The samples used in the high temperature tensile test are shown in Figure 1. At the clamping ends, two reinforcing pieces were stuck to both sides of the specimen in case of clamping failure during the experiment. Tensile tests were performed in the Instron 1332 equipment. A four-zone controlled high-temperature furnace was used. The temperature distribution on the gauge was well controlled within ±5 °C. The loading rate of tensile test was 2 mm/min. The heating rate was 30 °C /min and the holding time was 20 min. Five specimens were tested for each state and the strength was obtained by averaging the values over these five results.

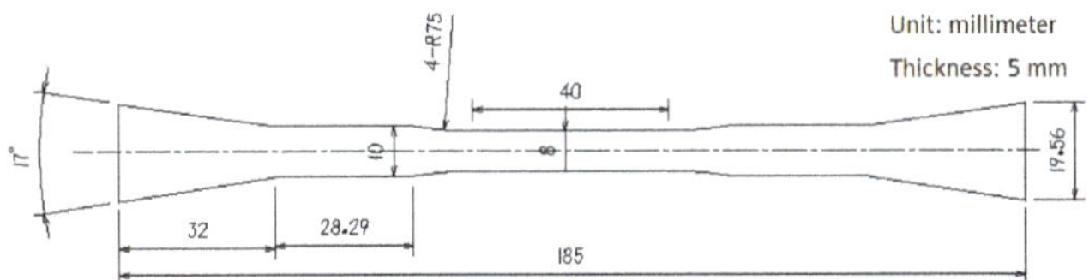

Figure 1. Sketch of the tensile test specimen.

3. Results and Discussion

3.1. Tensile Strength of SiC–SiC Composites at High Temperature in Air

The tensile strength of SiC–SiC composite was tested in situ at room temperature (RT), 800, 1200, 1300 and 1500 °C, respectively. The stress–strain curves at different temperatures are shown in Figure 2a. Each stress–strain curve is taken from the results with closet values to the average of all the five specimens at each temperature. The variations of tensile strength with temperature are summarized in Figure 2b. It can be seen from the results that the stress–strain curve barely changes at 800 °C. However, when the temperature further increases to 1200, 1300 and 1500 °C, the tensile strength decreases significantly by 48%, 58% and 66%, respectively. The tensile modulus slightly decreases with temperature from RT to 1300 °C. There is an obvious drop of tensile modulus at 1500 °C. The tensile modulus of composite is mainly determined by the modulus of fibers according to the composite material mechanics. When the temperature exceeds the preparation temperature of domestic Hi–Nicalon type SiC fibers (~1350 °C), the fiber modulus begins to degrade, leading to the obvious drop of SiC–SiC tensile modulus at 1500 °C.

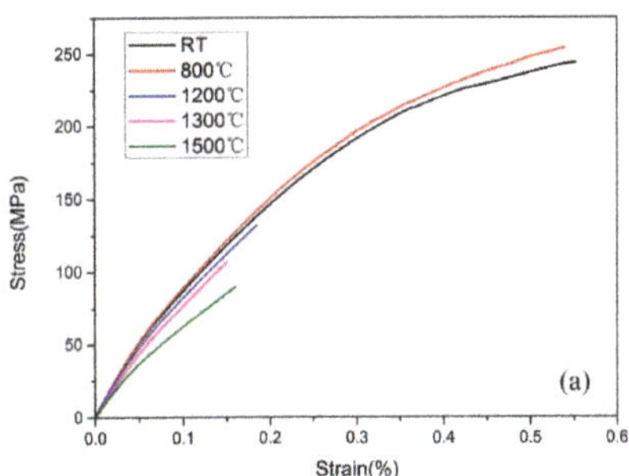
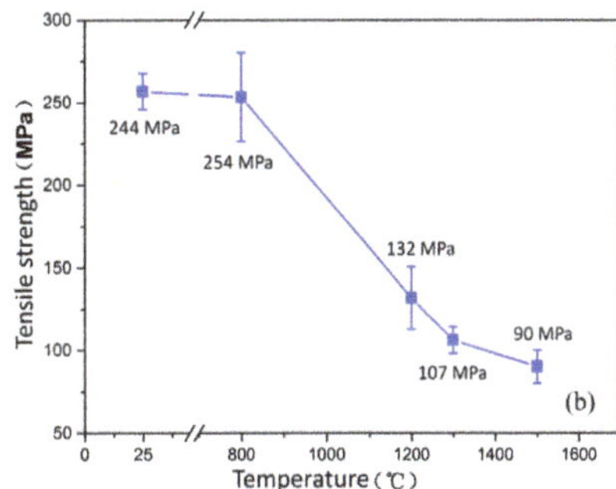

Figure 2. (**a**) Tensile stress–strain curves and (**b**) variations of tensile strength of SiC–SiC at different temperatures in air.

The microstructures at the fracture of SiC–SiC composites were characterized by Camscan Apollo 300 scanning electron microscope (CamScan, Cambridge, UK). The fractures of SiC–SiC composite at 800 °C exhibit ductile characteristics as shown in Figure 3. A large number of pulled out fibers with length of several millimeters can be seen at the fracture. When the temperature reaches 1200 °C, the length of pulling out fiber decreases to about tens of microns. When the temperature further increases to 1300 °C, the length of pulling out fibers further decreases and the regions of pulling out fibers become smaller. Most of the fibers are not pulled out, but break along with the matrix. When the temperature reaches 1500 °C, the fractures show typical brittle characteristics. The fracture surface of the composites is very plane and no pullout fibers can be seen. Cracks are not arrested at the interface, but penetrate the fiber bundles at 1500 °C.

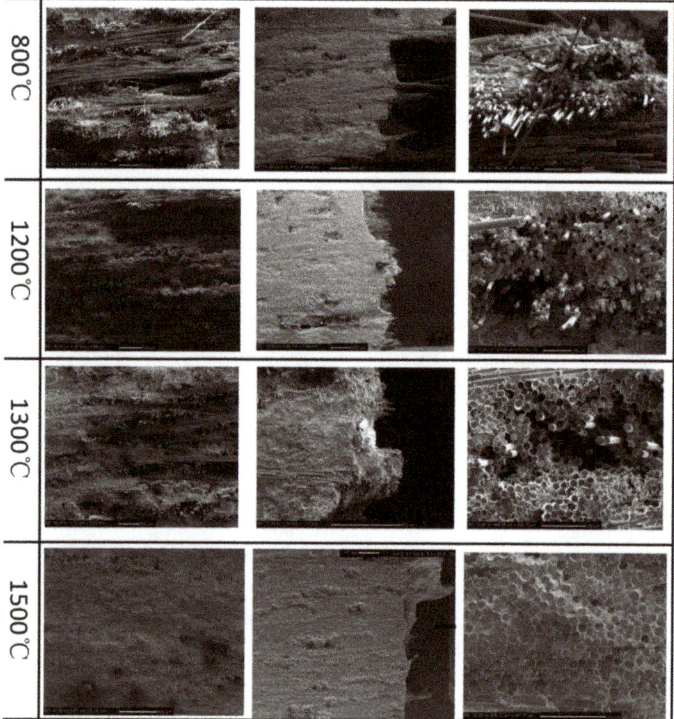

Figure 3. Tensile fracture morphology of SiC–SiC composites at high temperature in air.

3.2. Tensile Strength of SiC Fiber at High Temperature

In order to investigate the decline of tensile strength of SiC–SiC composites at high temperature, the tensile strength of monofilament SiC fiber at 1300 °C in air was tested in situ on the fiber ultra-high temperature performance testing system in Aerospace Research Institute of Materials & Processing Technology (Beijing, China). The gauge length was 40 mm. Monofilament SiC fiber was heated to 1300 °C with a heating rate of 40 °C/min and annealed for 5 min. The loading rate of tensile testing was 4 μm/s. The results are shown in Table 2. It can be seen that SiC fibers still retain high strength at 1300 °C and the strength retention rate is 78.9%, which is close to that of Hi–Nicalon fibers [17]. It indicates that the degradation of fiber is not the key factor that causes the sharp reduction of tensile strength of SiC–SiC composites at high temperature.

Table 2. Tensile strength of the SiC fiber at high temperature.

Room Temperature		1300 °C in Air		Strength Retention Rate (%)
Strength (GPa)	Cv Value (%)	Strength (GPa)	Cv Value (%)	
2.8	11.9	2.21	23.6	78.9

3.3. Influence of Fiber–Matrix Interface Coating on Tensile Properties of SiC–SiC Composites

The interface is known to play an important role in mechanical properties of materials [8,10,18,19]. According to the theory of fracture mechanics, the existence of fiber in the composite inhibits the crack propagation and improves the toughness of the material. In this study, it is assumed that both SiC fiber and SiC matrix is linear elasticity, and the relationship between interface shear stress and shear displacement is also linear. The contribution of fibers to the fracture toughness of composites is expressed as ΔK. Based on the fracture mechanics theory, the relationship between ΔK and interface strength τ_b is shown in formula (1) [20]

$$\Delta K = \begin{cases} \sqrt{\dfrac{d(E_mA_m+E_fA_f)\tau_b}{2(E_mA_mE_fA_f)}} \dfrac{E_fA_f\delta_b^{3/2}}{(A_f+A_m)\eta K_{IC}} \dfrac{\alpha}{\sqrt{1-a^2}}(\pi-2arcsin\alpha) & \left(\tau_b < \tau_b^C, \text{ pull out}\right) \\[2em] \sqrt{\dfrac{2(E_mA_mE_fA_f)\delta_b}{d(E_mA_m+E_fA_f)\tau_b}} \dfrac{(\sigma_f^b)A_f}{(A_f+A_m)\eta K_{IC}E_f} & \left(\tau_b > \tau_b^C, \text{ break}\right) \end{cases} \tag{1}$$

where, d is the diameter of the fiber, E_m and E_f are the Young's modulus of the matrix and fiber, A_m and A_f are the equivalent cross section areas of the matrix and fiber: $A_f = \pi d^2$, $A_m = d^2 - \pi d^2$, σ_f^b is the tensile strength of the fiber, τ_b is the interface strength, $\eta = 2\sqrt{2}(1-v_m^2)/(E_m\sqrt{\pi})$, where v_m is the Poisson's ratio of the matrix [21], δ_b is the critical interface shear displacement [22], K_{IC} and is the fracture toughness of the matrix. $\alpha = E_mA_m/(E_fA_f)$. τ_b^C Is the critical interface strength of fiber pull-out/break transformation, as shown in formula (2):

$$\tau_b^C = \frac{(\sigma_f^b A_f)^2}{CE_m^2A_m^2\delta_b} \tag{2}$$

$$C = \pi d \left(\frac{1}{E_fA_f} + \frac{1}{E_mA_m} \right) \tag{3}$$

The parameters involved in the formula above are shown in Table 3. We investigated the mechanical properties of SiC fiber and SiC matrix by nano-indentation tests. The results reveal that the modulus of SiC matrix is close to that of SiC fiber. For this reason, in this study, we assume that the SiC matrix has the same mechanical properties as SiC fiber for simplicity.

Table 3. Physical parameters used in fracture toughness calculation [21,22].

d	E_f	E_m	v_m	σ_f	δ_b
μm	GPa	GPa	-	GPa	μm
14	270	270	0.15	3	2

By substituting the data in Table 3 into Formulas (1) and (2), the relation of fracture toughness enhancement ΔK as a function of fiber–matrix interface strength can be obtained, as shown in Figure 4.

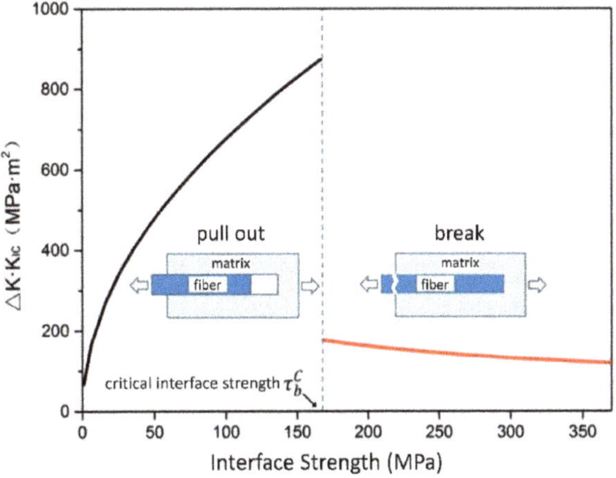

Figure 4. Fracture toughness enhancement of SiC–SiC as a function of fiber–matrix interface strength.

It can be seen from Figure 4 that when the failure mode is fiber pull-out, the toughening effect of the fiber increases with an increase in interface strength. Once the interface strength exceeds the critical value of fiber pull-out/break transition, the fracture toughness of SiC–SiC will drop suddenly. With the fiber-break failure mode, the fracture toughness of SiC–SiC decreases gradually with an increase in interface strength. For this reason, it is necessary to avoid the fiber-break failure mode for SiC–SiC composites. If SiC–SiC composites break in the fiber pull-out failure mode, further enhancement of the interface strength will improve the toughness of the material. Conversely, if SiC–SiC composites break in the fiber-break failure mode, the reduction of interface strength will improve the toughness of the material.

In SiC–SiC composites, the thermal expansion coefficient of the PyC interface coating is different from those of SiC fiber and SiC matrix. Therefore, the variation of temperature during PIP process will cause residual stress in the composites. Based on the characteristics of SiC–SiC composites, a two-dimensional finite element model of SiC–SiC unit cell is established, as shown in Figure 5. The model is composed of SiC fiber, PyC interface coating and SiC matrix. The fiber volume fraction is 40%. The material parameters used in the calculation are shown in Table 4. It is assumed that SiC fiber and SiC matrix are isotropic materials. The PyC interface coating is set to be transversal isotropy. The Young's modulus, sheer modulus, Poisson's ratio and coefficient of thermal expansion of PyC are determined according to the reference [23]. The Young's modulus of fiber is determined by our test. The Poisson's ratio of fiber is determined according to the reference [21]. The coefficient of thermal expansion of SiC fiber is determined by our test. We assume that the SiC matrix have the same mechanical properties as the SiC fiber. The densification process of PIP SiC matrix during heating above preparation temperature (850 °C) is not considered in this calculation. The thickness of the interface coating is 850 nm. The elements used are four-node bilinear plane stress elements (CPS4R), and the calculation is performed by Abaqus software. The simulation is divided into three analysis steps: 1. Set the initial analysis step. The initial temperature is set to 850 °C (SiC–SiC composite preparation temperature in this study). 2. Simulate the cooling process of SiC–SiC composites to room temperature (temperature field decreases from 850 to 25 °C). 3. Simulate the heating process tensile test of SiC–SiC composite to different temperatures and calculate the residual stress.

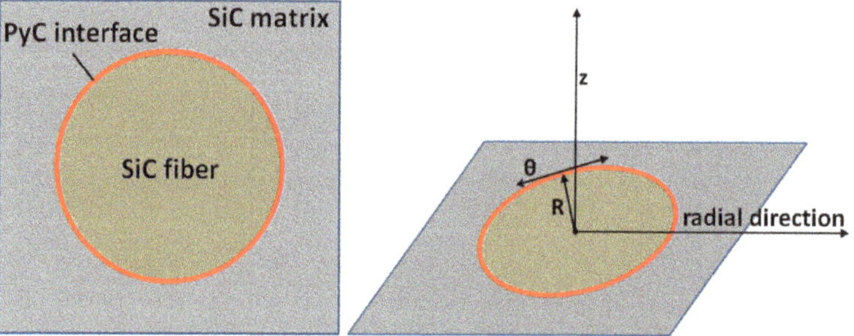

Figure 5. Sketch of SiC–SiC FEM model.

Table 4. Mechanical parameters used in the finite element calculation of residual stress [21,23].

Material	Young's Modulus (GPa)		Poisson's Ratio		Coefficient of Thermal Expansion (10^{-6} °C^{-1})	
	E_{zz}	E_{RR}	$V_{R\theta}$	V_{Rz}	α_{zz}	α_{RR}
SiC fiber	270	270	0.15	0.15	3.5	3.5
PyC interface	30	12	0.12	0.4	2	28
SiC matrix	270	270	0.15	0.15	3.5	3.5

During the tensile process, crack initiations in SiC–SiC composites mainly appear in the matrix and gradually propagate to the fiber–matrix interface coating. Experiments revealed that the failure mode of SiC–SiC composites is controlled by the interface strength [10]. If the fiber–matrix interface coating is weak enough, the crack will deflect into the interface and propagate along fiber axis direction. If the interface coating is too strong, the crack will not deflect, but directly propagate through the fibers. According to the Mohr–Coulomb criterion, shear strength is negatively correlated with the compressive stress perpendicular to the shear plane [24]. Therefore, the residual radial stress at the interface, σ_{RR}, has an important effect on the interface strength. The distribution of residual radial stress in SiC–SiC composites is shown in Figure 6.

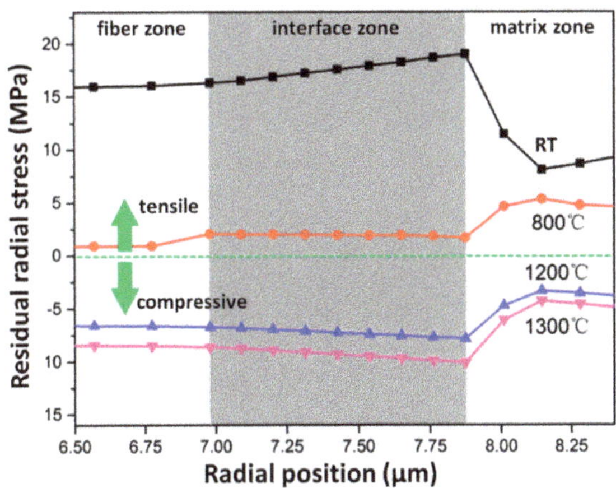

Figure 6. Distribution of residual radial stress, σ_{RR} of SiC–SiC at different temperature.

It can be seen clearly that the residual radial stress decreases with temperature. When the temperature exceeds 850 °C, the residual radial stress transforms from tensile to compressive, which prevents the crack propagation into the interface coatings. Therefore, at low temperature, the interface of SiC–SiC composite is relatively weak. Cracks tend to be deflected at the interface coatings, leading to fiber pull-out failure mode. With increase of temperature, the residual radial stress at the interface coatings decreases, which transforms from tensile to compressive. The interface becomes stronger at high temperature. Cracks can hardly be deflected and penetrate the fiber directly, leading to fiber-break failure mode and sudden reduction of tensile strength of SiC–SiC composite (as shown in Figure 7). Thus, one can attribute the transition from ductile to brittle at high temperature to the increase of interface strength.

The variations of residual hoop stress, $\sigma_{\theta\theta}$, with increasing temperature are shown in Figure 8. It can be seen from the results that below the preparation temperature, the residual hoop stress in the fiber and interface coatings is tensile. On the contrary, the residual hoop stress in the matrix is compressive. The residual hoop stress decreases with temperature. When the temperature exceeds preparation temperature, the residual hoop stress in the fiber and interface coatings converts to compressive and the residual hoop stress in the matrix converts to tensile. Above 850 °C, the residual hoop stress increases with temperature. According to the theory of fracture mechanics, compressive hoop stress in the interface coatings will restrain cracks extension. Therefore, the compressive hoop stress at high temperature will further enhance the interface and lead to tensile strength decline of the composites.

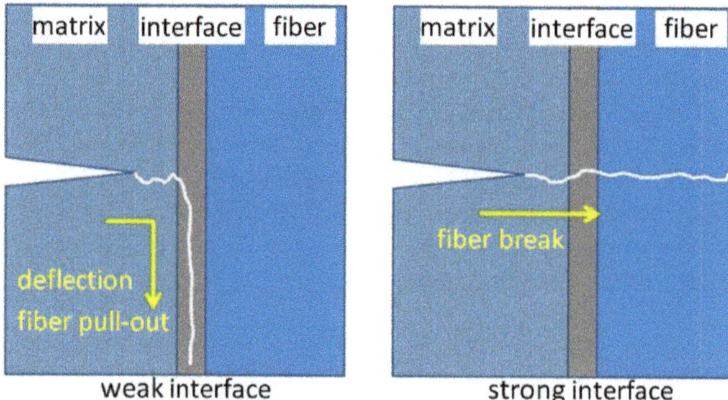

Figure 7. Schematic illustration of influence of interface strength on fracture mode of SiC–SiC.

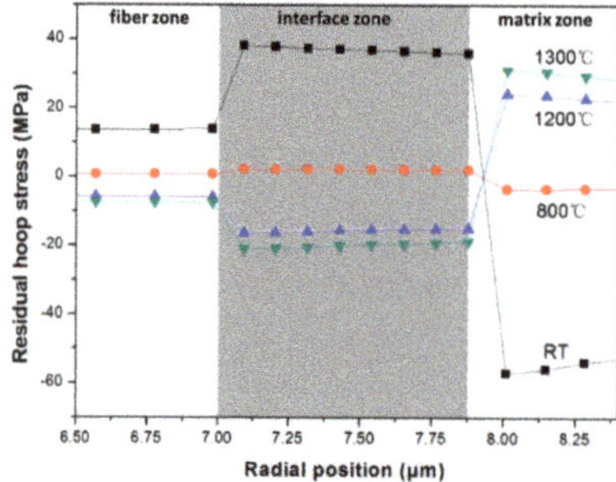

Figure 8. Distribution of residual hoop stress, $\sigma_{\theta\theta}$, of SiC–SiC at different temperature.

It is worthy to mention that when the temperature exceeds the preparation temperature (850 °C), the matrix will shrink due to PCS pyrolysis. This process will cause weight loss and crack initiations in the matrix, leading to acceleration of SiC–SiC strength reduction [10]. The chemical interactions between composite and air can also affect the mechanical properties of SiC–SiC at high temperature [11,12]. SiC matrix would form a glass phase at high temperature, which wraps the fiber and enhance the interface bonding [25]. The interface strength changes from weak to strong with increasing temperature by all factors discussed above and the composite changes from ductile fracture behavior to brittle fracture behavior.

We investigated the effect of interface coating thickness on the residual radial stress of SiC–SiC composites at 1300 °C. It can be seen from Figure 9 that at this temperature, the residual radial stress at the interface coating is compressive and decreases with an increase in interface thickness. The SiC–SiC composite breaks in fiber-break failure mode at 1300 °C. Therefore, reducing the thickness of the interface can reduce the interface strength and improve the fracture toughness of the SiC–SiC composites according to Figure 4.

To verify our theory, SiC–SiC composites with different interface thicknesses are prepared. The microstructures of SiC–SiC composites with average interface thickness of 100, 360, 580, and 640 nm are shown in Figure 10a. The tensile strength of SiC–SiC composites with different interface thicknesses at 1300 °C in air are shown in Figure 10b. It can be seen from the results that the tensile strength of SiC–SiC composite increases with decreasing interface thickness, in good agreement with the theoretical analysis above. It is necessary to mention that if there is no PyC interface coating, the SiC fiber and SiC matrix will sinter together at high preparation temperature. There will be strong interactions between fiber–matrix and the mechanical strength of the SiC-SiC composite is very low.

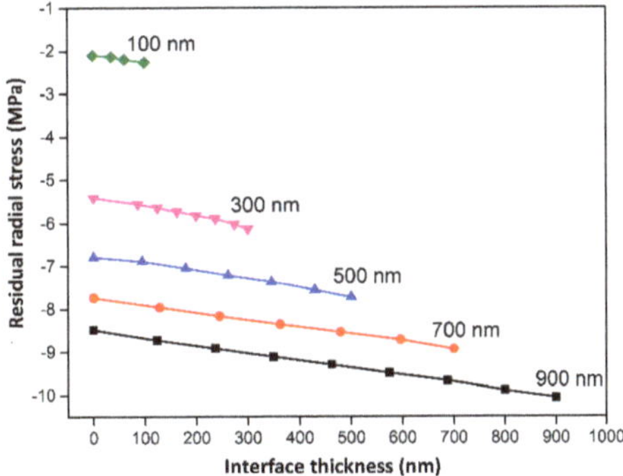

Figure 9. Distribution of residual stress, σ_{RR} at 1300 °C with different interface thickness.

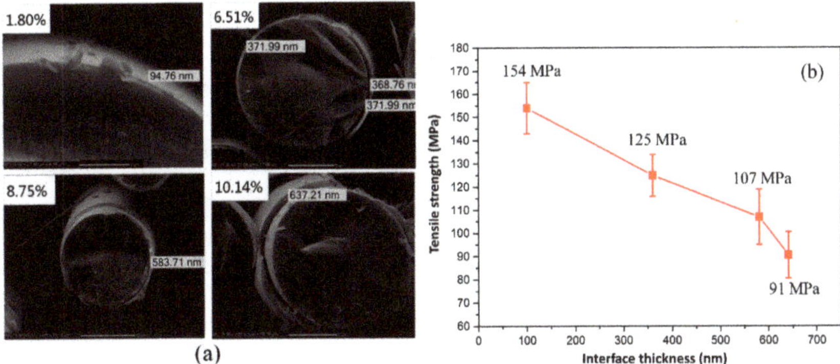

Figure 10. (**a**) SEM images of SiC–SiC composites with different interface thickness; (**b**) tensile strength of SiC–SiC composites with different interface thicknesses at 1300 °C in air.

4. Conclusions

- The tensile properties of SiC–SiC composites reinforced by domestic Hi–Nicalon type SiC fibers at 800, 1200, 1300, and 1500 °C were studied. The results show that when the temperature exceeded 1200 °C, the tensile strength of SiC–SiC composites decreased sharply and the composite failure mode converted from fiber-pull-out to fiber-break.
- Theoretical calculations showed that the toughening effect of fibers increases with interface strength when the composite failure mode is fiber-pull-out. Once the interface strength exceeds the critical value, the composite failure mode converts to fiber-break and the fracture toughness of SiC–SiC drops sharply. The finite element method simulations show that when the temperature exceeds the material preparation temperature, the residual radial stress at the interface increases and changes from tensile to compressive, causing transition of the failure mode and sudden reduction of tensile strength.
- Theoretical and experimental results showed that reducing the thickness of PyC interface coating improved the tensile strength of SiC–SiC composites at high temperatures.

Author Contributions: Conceptualization, E.J., X.S., Z.F., and J.L.; Data curation, W.S., H.L., D.M., and X.S.; Formal analysis, H.L., K.W., and J.L.; Funding acquisition, D.M. and Z.F.; Investigation, E.J., W.S., X.S., and J.L.; Methodology, E.J. and K.W.; Project administration, Z.F.; Resources, D.M.; Software, K.W.; Writing – original draft, E.J.; Writing – review & editing, W.S., H.L., X.S., J.L., and Z.Y. All authors have read and agreed to the published version of the manuscript.

Funding: This research received no external funding.

Acknowledgments: The authors thank Hongmei Yang, Lina Huang and Xinpeng Wang for outstanding work during experiments.

Conflicts of Interest: The authors declare no conflict of interest.

References

1. Wang, P.; Liu, F.; Wang, H.; Li, H.; Gou, Y. A review of third generation SiC fibers and SiCf/SiC composites. *J. Mater. Sci. Technol.* **2019**, *35*, 2743–2750. [CrossRef]
2. Zhai, Z.; Wei, C.; Zhang, Y.; Cui, Y.; Zeng, Q. Investigations on the oxidation phenomenon of SiC/SiC fabricated by high repetition frequency femtosecond laser. *Appl. Surf. Sci.* **2020**, *502*, 144131. [CrossRef]
3. Raj, S. Development and characterization of hot-pressed matrices for engineered ceramic matrix composites (E-CMCs). *Ceram. Int.* **2019**, *45*, 3608–3619. [CrossRef]
4. Koyanagi, T.; Katoh, Y.; Nozawa, T.; Snead, L.; Kondo, S.; Henager, C.; Ferraris, M.; Hinoki, T.; Huang, Q. Recent progress in the development of SiC composites for nuclear fusion applications. *J. Nucl. Mater.* **2018**, *511*, 544–555. [CrossRef]
5. Xu, S.; Li, X.; Zhao, Y.; Liu, C.; Mao, Q.; Cheng, L.; Zhang, L. Micromechanical properties and microstructural evolution of Amosic-3 SiC/SiC composites irradiated by silicon ions. *J. Eur. Ceram. Soc.* **2020**, *40*, 2811–2820. [CrossRef]
6. Braun, J.; Sauder, C.; Lamon, J.; Balbaud-Célérier, F. Influence of an original manufacturing process on the properties and microstructure of SiC/SiC tubular composites. *Compos. Part A* **2019**, *123*, 170–179. [CrossRef]
7. Wang, X.; Song, Z.; Cheng, Z.; Han, D.; Li, M.; Zhang, C. Tensile creep properties and damage mechanisms of 2D-SiC$_f$/SiC composites reinforced with low-oxygen high-carbon type SiC fiber. *J. Eur. Ceram. Soc.* **2020**, in press. [CrossRef]
8. Buet, E.; Sauder, C.; Sornin, D.; Poissonnet, S.; Rouzaud, J.; Vix-Guterl, C. Influence of surface fibre properties and textural organization of a pyrocarbon interphase on the interfacial shear stress of SiC/SiC minicomposites reinforced with Hi-Nicalon S and Tyranno SA3 fibres. *J. Eur. Ceram. Soc.* **2014**, *34*, 179–188. [CrossRef]
9. Buet, E.; Sauder, C.; Poissonnet, S.; Brender, P.; Gadiou, R.; Vix-Guterl, C. Influence of chemical and physical properties of the last generation of silicon carbide fibres on the mechanical behaviour of SiC/SiC composite. *J. Eur. Ceram. Soc.* **2012**, *32*, 547–557. [CrossRef]
10. Fellah, C.; Brauna, J.; Sauder, C.; Sirotti, F.; Berger, M. Influence of the carbon interface on the mechanical behavior of SiC/SiC composites. *Compos. Part A* **2020**, *133*, 105867. [CrossRef]
11. Hsu, C.-Y.; Zhang, Y.; Xie, Y.; Deng, F.; Karandikar, P.; Xiao, J.Q.; Ni, C.; Xie, Y.; Karandikar, P. In-situ measurement of SiC/Si interfacial tensile strength of reaction bonded SiC/Si composite. *Compos. Part B* **2019**, *175*, 107116. [CrossRef]
12. Ding, D.; Zhou, W.; Luo, F.; Chen, M.; Zhu, D. Mechanical properties and oxidation resistance of SiC$_f$/CVI-SiC composites with PIP-SiC interphase. *Ceram. Int.* **2012**, *38*, 3929–3934. [CrossRef]
13. Wang, L.-Y.; Luo, R.-Y.; Cui, G.-Y.; Chen, Z.-F. Oxidation resistance of SiC$_f$/SiC composites with a PyC/SiC multilayer interface at 500 °C to 1100 °C. *Corros. Sci.* **2020**, *167*, 108522. [CrossRef]
14. Shimoda, K.; Hinoki, T.; Park, Y.-H. Development of non-brittle fracture in SiC$_f$/SiC composites without a fiber/matrix interface due to the porous structure of the matrix. *Compos. Part A* **2018**, *115*, 397–404. [CrossRef]
15. Kato, Y.; Ozawa, K.; Shih, C.; Nozawa, T.; Shinavski, R.J.; Hasegawa, A.; Snead, L.L. Continuous SiC fiber, CVI SiC matrix composites for nuclear applications: Properties and irradiation effects. *J. Nucl. Mater.* **2014**, *448*, 448–476. [CrossRef]
16. Lamon, J. Properties of Characteristics of SiC and SiC/SiC Composites. *Comp. Nucl. Mater.* **2012**, *2*, 323–338.
17. Sha, J.; Hinoki, T.; Kohyama, A. Thermal and mechanical stabilities of Hi-Nicalon SiC fiber under annealing and creep in various oxygen partial pressures. *Corros. Sci.* **2008**, *50*, 3132–3138. [CrossRef]
18. Chen, P.; Chen, S.; Peng, J.; Gao, F.; Liu, H. The interface behavior of a thin film bonded imperfectly to a finite thickness gradient substrate. *Eng. Fract. Mech.* **2019**, *217*, 106529. [CrossRef]
19. Chen, P.; Chen, S.; Peng, J. Interface behavior of a thin-film bonded to a graded layer coated elastic half-plane. *Int. J. Mech. Sci.* **2016**, *115*, 489–500. [CrossRef]
20. Chen, Y.; Liu, B.; He, X.; Huang, Y.; Hwang, K. Failure analysis and the optimal composites design of carbon nanotube-composites. *Compos. Sci. Tech.* **2010**, *70*, 1360–1367. [CrossRef]

21. Drissi-Habti, M.; Nakano, K.; Kagawa, Y. Modelling of the interfacial behaviour in the Hi-Nicalon fibre-reinforced α-Si$_3$N$_4$ ceramic matrix composites using microindentation tests. *Mat. Sci. Eng. A* **1998**, *250*, 178–185. [CrossRef]
22. Mueller, W.; Moosburg-Will, J.; Sause, M.; Horn, S. The Microscopic analysis of single—fiber push—out tests on ceramic matrix composites performed with Berkovich and flat-end indenter and evaluation of interfacial fracture toughness. *J. Eur. Ceram. Soc.* **2013**, *33*, 441–451. [CrossRef]
23. Li, J.; Lv, Z.; Ren, C.; Chen, G.; Dong, J. Finite element simulation of interface shear strength of c/sic composites with thermal residual stress. *J. Mat. Sci. Eng.* **2017**, *35*, 801–805. (In Chinese)
24. Wang, S.F.; Feng, J.L.; Yang, Z.J. Two active plane finite element model in Mohr-Coulomb elastoplasticity. *Finite Elem. Anal. Des.* **1999**, *32*, 213–219. [CrossRef]
25. Chai, Y.; Zhou, X.; Zhang, H. Effect of oxidation treatment on KD-II SiC fiber–reinforced SiC composites. *Ceram. Int.* **2017**, *43*, 9934–9940. [CrossRef]

Article

Research on Interface Characteristics and Mechanical Properties of 6061 Al Alloy and Q235a Steel by Hot Melt-Explosive Compression Bonding

Yu-ling Sun [1,2], Hong-hao Ma [1,3,*], Ming Yang [1], Zhao-wu Shen [1], Ning Luo [4,5,*] and Lu-qing Wang [1]

1   CAS Key Laboratory of Mechanical Behavior and Design of Materials,
    University of Science and Technology of China, Hefei 230027, China; ylsun16@mail.ustc.edu.cn (Y.-l.S.);
    ym1991@ustc.edu.cn (M.Y.); ZWShen@ustc.edu.cn (Z.-w.S.); aiyuan@ustc.edu.cn (L.-q.W.)
2   High Overloaded Ammunition Guidance Control and Information Perception Laboratory,
    PLA Army Academy of Artillery and Air Defense, Hefei 230031, China
3   State Key Laboratory of Fire Science, University of Science and Technology of China, Hefei 230027, China
4   State Key Laboratory for Geomechanics and Deep Underground Engineering,
    China University of Mining and Technology, Xuzhou 221116, China
5   School of Mechanics and Civil Engineering, China University of Mining and Technology,
    Xuzhou 221116, China
*   Correspondence: hhma@ustc.edu.cn (H.-h.M.); nluo@cumt.edu.cn (N.L.)

Received: 22 September 2020; Accepted: 21 October 2020; Published: 26 October 2020

Abstract: In order to solve the shortcomings of the traditional explosion welding method for direct magnesia-aluminum alloy and steel welding, a processing method of groove hot casting plus explosion compression bonding (HCECB) was put forward, and the related theory of hot-melt metal plus explosion bonding was also proposed. Taking 6061 aluminum and Q235a steel as examples, the hot casting plus explosion compression test was carried out by the prefabrication of a dovetail groove on Q235a steel plate and the microstructure and mechanical properties of the interface were analyzed. The results showed that the 6061 aluminum/Q235a steel can be directly combined by the HCECB method. The interface is mainly irregularly microwave-shaped and straight-shaped with no defects, such as melting layer, holes and cracks, found. The hardness of the upper interface of the dovetail groove is larger than that far away from the welding interface, while the hardness of the lower interface is the same as that far away from the interface. The tensile and shear test results show that the shear strength is greater than 80 MPa, which meets the requirements of aluminum-steel composite plate bonding strength.

Keywords: hot casting metals; explosive compression bonding; dovetail groove; metallographic analysis; mechanics analysis

1. Introduction

Aluminum and steel laminated composite materials are widely used in various industrial fields because of their respective advantages, e.g., aluminum has excellent oxidation resistance, corrosion resistance and electrical and thermal conductivity, while steel has the advantages of high strength, hardness and wear resistance [1,2]. Explosive welding uses the energy of explosives to drive the high-speed collision of aluminum steel plate to realize metallurgical bonding. It is the main method to produce aluminum and steel composite materials [3,4]. Elango et al. studied the effect of variable load ratio (explosive mass/fin mass) on the properties, temperature and pressure of the explosive welded interface of aluminum steel [5]. In fact, it is difficult to realize the direct welding between aluminum alloy and steel due to the different physical and chemical properties between aluminum and steel.

Additionally, the interface between aluminum and steel easily to forms Fe_3Al, Fe_2Al_5, $FeAl_2$ and other brittle intermetallic compounds [6–9]. In recent years, the composite methods of aluminum and steel mainly include brazing, laser welding, argon arc welding, cold metal transfer welding, friction stir welding [10–20], etc. Singh P et al. made a comparative evaluation on the welding methods of the dissimilar metal joints between aluminum and stainless-steel [21].

In order to obtain more ideal welding results, researchers have widely used the method of adding a cushion between aluminum and steel interface. Hokamoto et al. [22] used thin stainless-steel plates as transition plates to reduce the brittleness of intermetallic compounds. Aceves et al. [23] studied the advantages and disadvantages of Ta, Cu, Ti and other metals as intermediate layers in the explosive welding of aluminum steel. The explosive welding between aluminum alloy and steel can be realized by using intermediate transition layer. However, its operation is complicated, and the introduction of an intermediate layer may reduce the application prospects of aluminum steel composite. Li et al. [24] studied the direct explosion of aluminum and stainless-steel with dovetail grooves. This is a very good method to make use of groove structure meshing, aluminum and stainless-steel to achieve the purpose of the explosive compound, but it may cause groove structure deformation.

In this paper, a new groove hot casting plus explosion compression bonding (HCECB) method is introduced, which provides a new idea for welding 6061 aluminum alloy with Q235a steel. The HCECB theory is based on the combination of hot-melt bonding and high impact explosive bonding, which is more conducive to the strong bonding between metals under thermal impact. For the whole HCECB process, first of all, the aluminum alloy solution was poured onto the surface of the etched dovetail groove, and then explosive compression was performed. The use of dovetail grooves can prolong the service life of aluminum stainless-steel composites. Then the hot-melt plate was squeezed into the dovetail groove by the HCECB method, which can protect the dovetail groove from deformation and realize the compact combination of aluminum and steel.

2. Experimental

2.1. The Experimental Process

The flyer plate and the base plate were made from aluminum (6061) and steel (Q235a), respectively. The chemical compositions of the two materials are shown in Tables 1 and 2. The experimental process is divided into two parts. In the experimental process, the dovetail grooves were firstly made on the surface of Q235a, and the interface of the grooves was cleaned. Then, 6061 aluminum was melted in a high temperature furnace. Part of the liquid aluminum flowed into the tank, the initial combination of 6061 and Q235a was achieved under the action of the heat of the high-temperature liquid and the mechanical action of the tank type. Then, the first step was to pour the molten aluminum onto the stainless-steel surface with dovetail groove, to achieve the preliminary hot-melt composites. The second step was to make the hot-melt flyer-plate further squeeze into the dovetail groove with the HCECB method, in order to achieve the strong association between the impact of thermal-mechanical metal. Finally, the explosive compression method is used to push the base plate and the cover plate to squeeze together, compacting and combining, forming a close bonding force at the interface of the two metal plates, to achieve the strong bonding between the metals under the heat-force impact.

Table 1. Chemical composition of 6061-Al (wt %).

Element	Cu	Mn	Mg	Zn	Cr	Ti	Si	Fe	Al
Content	0.15–0.4	≤0.15	0.8–1.2	≤0.25	0.04–0.35	≤0.15	0.4–0.8	≤0.7	Bal.

Table 2. Chemical composition of Q235a (wt %).

Element	C	Mn	Si	S	P	Fe
Content	0.14–0.22	0.30–0.65	≤0.30	≤0.05	≤0.045	Bal.

2.2. Experimental Parameters

2.2.1. Hot-Melt Composite Plate

Dovetail grooves with an upper bottom surface of 2 mm, a lower bottom surface of 3 mm and a height of 1 mm were cut out on the upper surface of the substrate, 1 mm from the spacing on the upper and bottom surface of the swallowtail groove. The heating furnace was used in the experiment and the maximum temperature can reach up to 1000 °C. In this experiment, the maximum melting temperature was 850 °C and it was kept for 1 h. The aluminum alloy was melted, and then poured on the surface of the stainless-steel dovetail groove to make the aluminum liquid fully flow into the stainless-steel dovetail groove, and then cooled naturally to form the 6061Al-Q235a steel hot-melt composite plate.

In this step, the thermal bonding refers to the mutual bonding of molten aluminum and stainless-steel. The liquid metal Al alloy is poured into the stainless-steel dovetail tank. Heat transfers on the surface of the stainless-steel. Heat energy is continuously transmitted to the stainless-steel. The Al alloy and stainless-steel achieve a hot fusion combination, due to the intensification of thermal motion inside the molecules. Thermal stress will be generated due to the transfer of heat energy from liquid aluminum to the stainless-steel plate. The temperature on the surface of the stainless-steel dovetail groove is higher than that on the inside. Thermal stress generally makes the junction surface shrink, which is not good for the composite of aluminum and stainless-steel. However, it is beneficial to the combination of aluminum and stainless-steel for the special dovetail groove structure. The generation of thermal stress will cause the aluminum in the dovetail groove to be squeezed up and down and store thermal stress. Aluminum and stainless-steel were preliminarily combined under the thermal action of the high temperature of the aluminum liquid and the mechanical action of the dovetail groove.

2.2.2. Explosive Compression Composite Plate

In order to make the interface between Al alloy and stainless-steel closer, the explosive compression is further carried out to achieve the strong bonding between aluminum and stainless-steel under the heat-force impact. After fusion and treatment, the size of the hot-melt composite plate is $165 \times 125 \times 24$ mm^3, and the thickness of aluminum is about 25 mm. Honeycomb explosives were used in the experiment, as shown in Figure 1. The selected explosive was emulsion explosive with a density of about 0.8 g/cm^3 and the explosive velocity was about 2600 m/s. We observe the conservation of energy in the welding process, and economized the use of explosives [25]. The charge thickness was 10 mm. The uniformity of charge thickness and density was ensured effectively.

The top of the explosive was covered with a colloidal water bag to increase the constraint and improve the utilization rate of the explosive in order to reduce the explosive energy flying upward (Figure 2). The parameters of the explosion compression are shown in Table 3. In order to investigate the microstructure of the joining interface, the test block of dimensions was cut from the welded plate parallel to the detonation direction. The cross-section of the block was polished using emery papers and diamond slurries. The microscopic examination of the composites was carried out using an optical microscope. In addition, higher-resolution examination and elemental analysis were investigated using a high-resolution field emission scanning electron microscope (SEM) (MAIA3 LMH, Hefei, China). Vickers microhardness test was performed using an HVS-1000A type microhardness testing machine (Hefei, China) with a load of 9.807 N and the loading time was 10 s.

Explosive compression bonding is used to achieve the strong bonding between 6061 aluminum and Q235a stainless-steel under heat-force impact, in order to make the interface between aluminum alloy and stainless-steel closer. Large stress is stored in dovetail grooves due to the impact of explosive energy, which causes the aluminum plate to produce plastic deformation, or even flow deformation. A tight bond was formed due to the Al alloy plate being further adiabatically compressed to the dovetail channel. At the same time, the inclined plane of the dovetail groove prevents the stretching wave from pulling the interface apart, realizing further recombination [24,26].

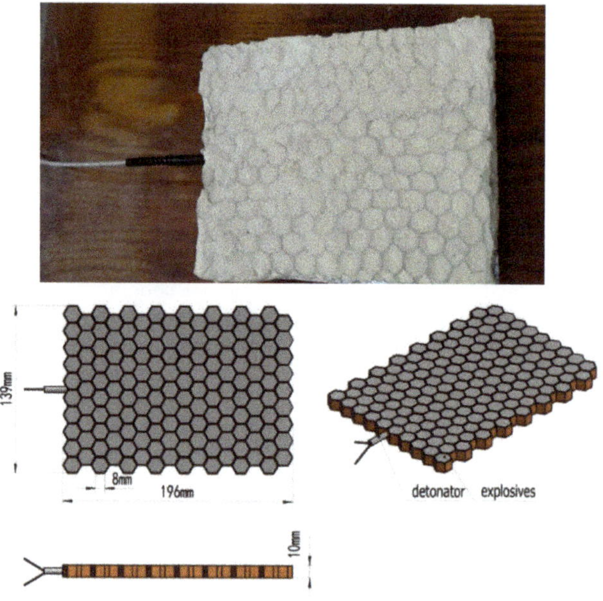

Figure 1. Honeycomb aluminum structure explosive.

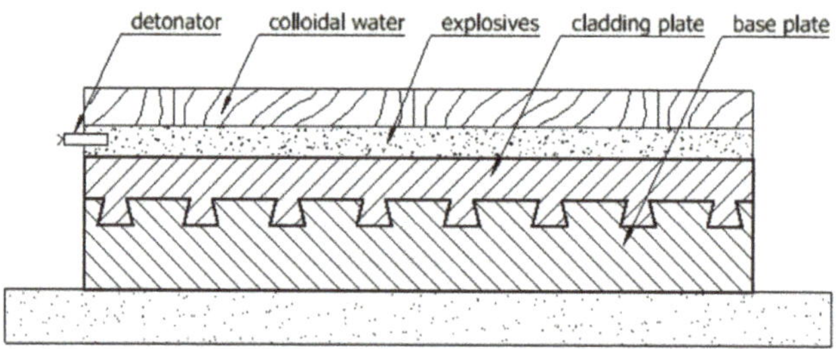

Figure 2. Schematic diagram of explosive compression structure device.

Table 3. Parameters of explosion compression bonding of 6061-Q235a.

Parameters	Density of Explosive/g·cm^{-3}	Charge Mass/g	Layout Thickness/mm	Charge at Unit Area/g·cm^{-2}	Mass Ratio
Parameter values	0.9	275	10	0.9	0.55

3. Experimental Results and Analysis

3.1. Welding Interface Structure Analysis

The interface of the 6061-Q235a composite plate was closely combined by observing the section along the direction of detonation after the explosion compression, and the dovetail groove has a clear structure without any damage marks, as shown in Figure 3.

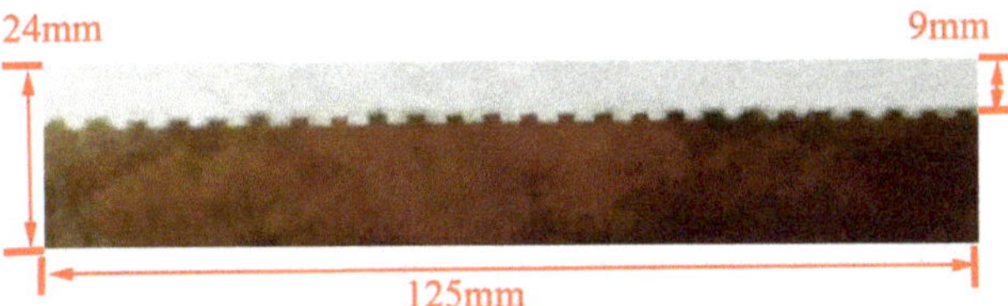

Figure 3. The physical cross-section of the hot-melt explosive composite plate.

The tensile shear experiment of the 6061 Al-Q235a composite plate was done subjected to the standard of GB/T 6386-2008 [27]. The obtained samples were taken, and the size was 10 mm × 1 mm × 45 mm. The mechanical properties and microstructure of the welding interface were tested by the MC010 touch screen digital microhardness tester (Hefei, China) and Carl Zeiss Axio Imager metallographic microscope (Hefei, China) with SEM and EDS analysis, respectively.

3.2. Observation of Interface Metallographic Structure

After polishing the 10 mm × 10 mm × 24 mm sample of the composite plate, the metallographic structure of the interface between aluminum and dovetail groove was observed by a metallographic microscope, as shown in Figure 4. The position of the interface between 6061 and Q235a groove steel obtained by the metallographic microscope is shown in Figure 4a. The metallographic structures of the upper, lower and lateral sides of 6061 Al alloy and Q235a groove steel are shown as Figure 4b–d, respectively. The metallographic photograph shows that the interface of the composite plate is combined in a straight and microwave manner. The wavelength of the interface is about 70–100 μm. The quality of explosive compression is good because the interface of explosive composite plate is closely combined, and there is basically no melt.

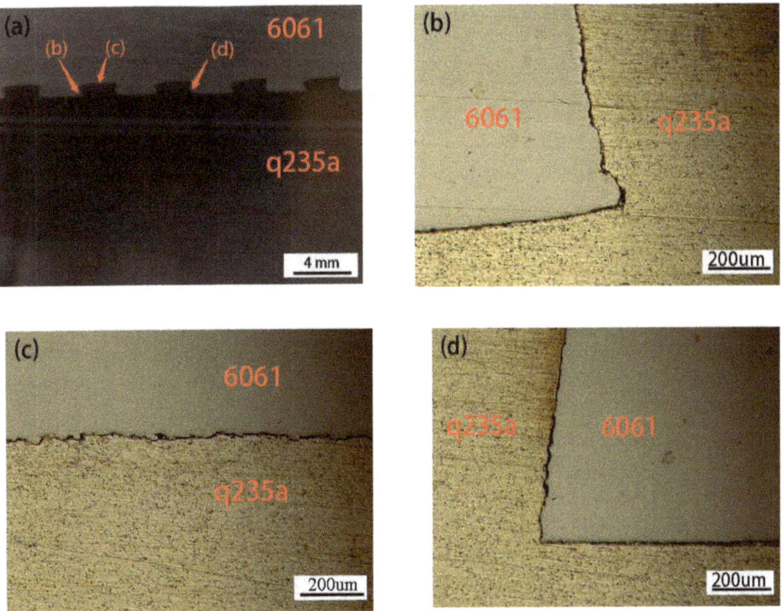

Figure 4. The different metallographic observation structure of groove hot casting plus explosion compression bonding (HCECB) composite plates: (**a**) Metallographic observation location map; (**b**) Bottom and side of dovetail groove; (**c**) the upper surface; (**d**) side and Bottom of dovetail groove.

3.3. Interface Structure Analysis

The microstructure of the 6061-Q235a HCECB composite plate was observed by SEM and EDS, and the chemical elements of the interface were analyzed. Figure 5 shows the SEM morphology and EDS delamination image of the 6061-Q235a hot-melt explosive composite plate interface. It can be seen from Figure 5 that the bonding interface is close, with mainly mechanical bonding, and no obvious melting zone is found, stainless-steel is on the left, and aluminum is on the right. It was found that the iron element penetrates into the Al alloy near the welding interface, and no obvious aluminum element penetrates into stainless-steel, this may be because the hardness of stainless-steel is greater than that of aluminum.

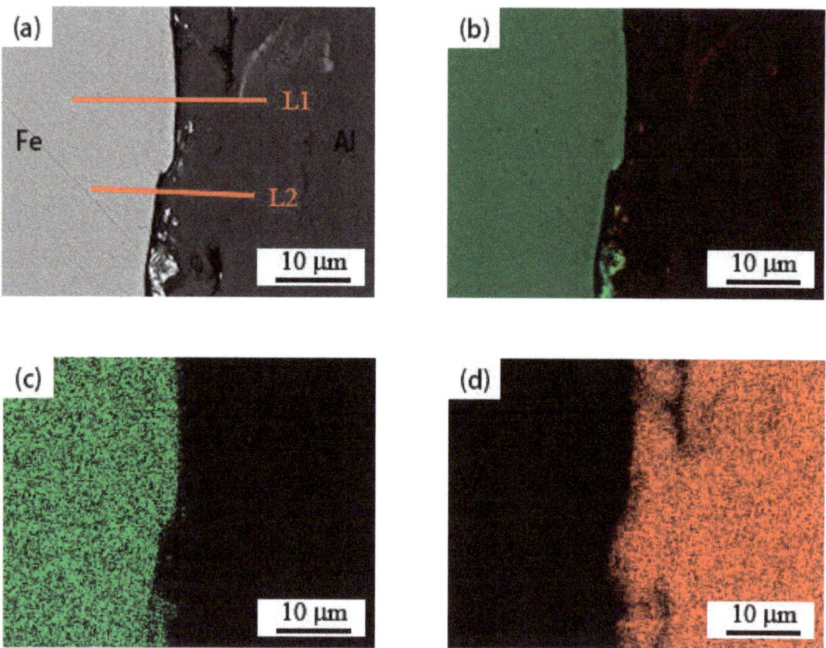

Figure 5. Elemental mapping results of the Al-6061/Q235a joining interface (**a**) SEM image of the joining interface; (**b**) elemental mappings of the joining interface; (**c**) the elemental distribution of Fe; (**d**) the elemental distribution of Al.

Fan et al. [28] also found that the Al/Fe diffusion interface was mainly caused by the rapid migration of Fe atoms in the disordered Al matrix in the Al bimetallic tube made by the magnetic pulse composite method. Figure 6 shows the SEM morphology and the line energy spectrum of the main elements at the interface of the 6061-Q235a hot-melt explosive composite plate. The location of the line sweep is shown in the first figure in Figure 5. It can be seen that there is a mutual diffusion between atoms at the interface junction, with aluminum and iron accounting for about half of each, which shows that the 6061 and Q235a interface is well combined, resulting in the mutual penetration of elements. At the same time, the diffusion of iron to aluminum was also found, but aluminum did not show any obvious diffusion to iron. From the SEM results, the interface of the 6061-Q235a hot-melt explosive composite plate, the surface energy spectrum and the line energy spectrum of the main elements, it can be seen that the interface of the composite plate is dense and the bonding quality is good.

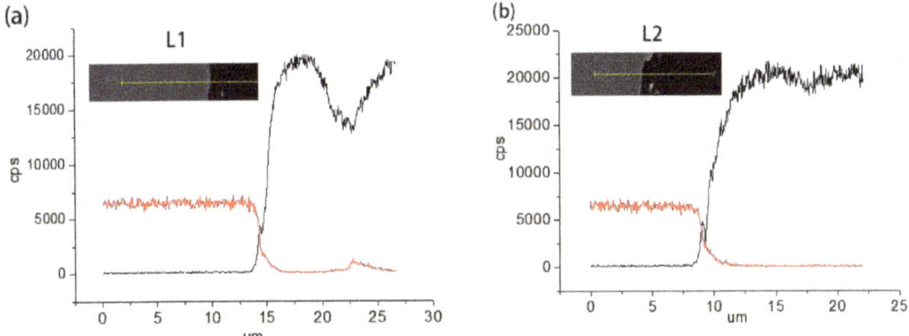

Figure 6. Linear energy spectra of the main elements of the welding interface in two different areas
(**a**) area of L1; (**b**) area of L2.

3.4. Microhardness Analysis

The microhardness was measured after the 10 mm × 10 mm × 24 mm sample of the composite plate was polished. The microhardness of metal on both sides of the composite plate interface of the vertical sample interface was measured by the MC010 touch screen digital microhardness tester. The microhardness near the interface of the bottom and top of the dovetail groove was measured. The test position spacing was 0.2 mm, 0.200 kgf, 15 s. The hardness test position and test results of the composite board are shown in Figures 7 and 8, respectively.

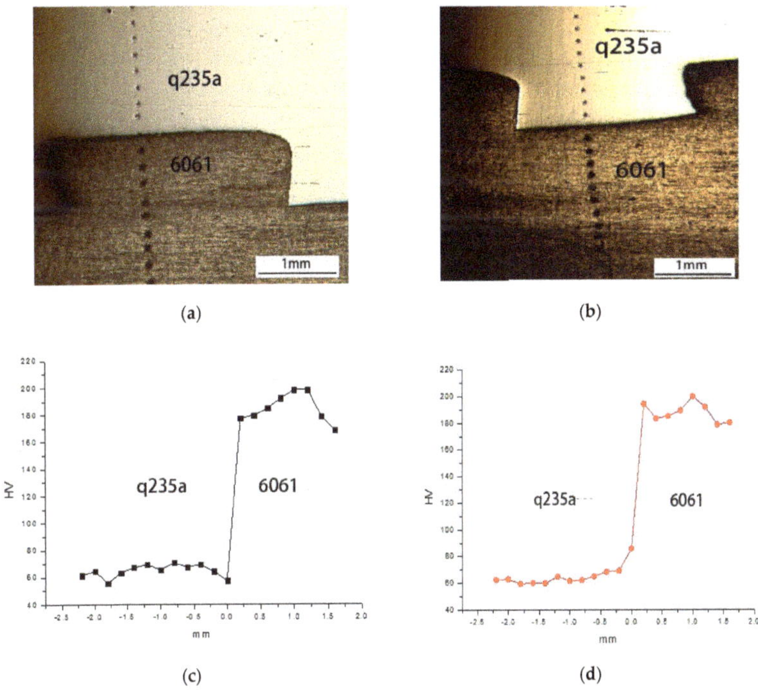

Figure 7. The microhardness characteristics and corresponding results of two different areas at the joining interface: (**a**) Test position of microhardness (the upper of dovetail groove); (**b**) Test position of microhardness (the bottom of dovetail groove); (**c**) test results of microhardness (the upper of dovetail groove); (**d**) test results of microhardness (the bottom of dovetail groove).

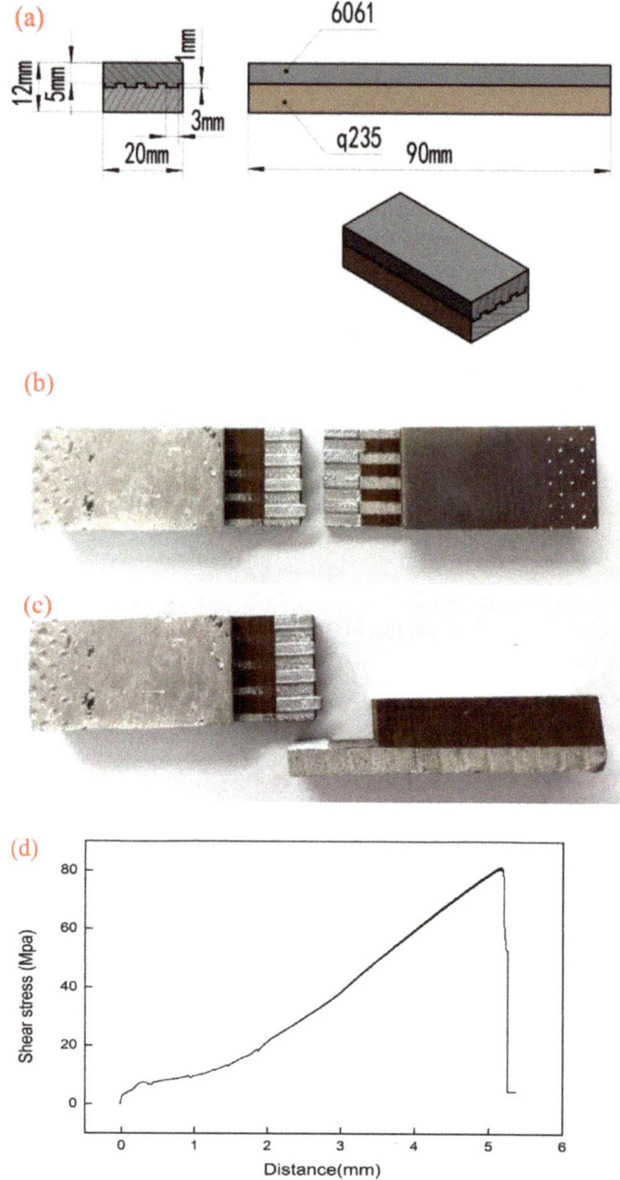

Figure 8. The sample diagrams and results of the shear force test. (**a**) The sketch diagram of shear force test sample (**b,c**) The samples of Al-6061/Q235a composite plate for the shear force test; (**d**) shear force curve of the composite plate.

As can be seen from Figure 7, the bonding strength of the composite surface under the dovetail groove is smaller than that of the upper surface. At this interface, the Vickers hardness of aluminum (hv57-69) is greater than the minimum value of Vickers hardness of aluminum 6061 (hv44-95), the microhardness of Q235a near the interface is about 177HV. The microhardness of 6061 and Q235a near the composite interface of the upper and bottom surface of dovetail groove had increased.

The microhardness at the interface is 85HV, the microhardness of stainless-steel on the side of stainless-steel 0.2 mm away from the interface reaches 194HV, which is close to the maximum value. When the distance from the interface is about 1.0 mm, the microhardness of the upper and lower surface reaches the maximum. This is because, as seen in the binary diagram of iron and aluminum, Al has a certain solubility in Fe, while Fe is almost insoluble in A1. A1 diffuses into Fe, producing many kinds of intermetallic compound, such as Fe_3Al, $FeAl$, $FeAl_2$, $FeAl_5$ and $FeAl_3$. The microhardness of composite plates increases because of these intermetallic compounds. Near the Al side, because Fe is almost insoluble in A1, the microhardness is less affected. On the whole, the microhardness of the composite interface and its vicinity conforms to the microhardness requirements of the composite materials.

3.5. Tension Shear Test

The interface bonding strength is one of the most important indexes to evaluate the quality of explosive welding. Shear strength is used to measure the mechanical properties of the explosive composite plate, referring to the GB/T 6386-2008 test method [27] for mechanical and technological properties of composite steel plates. The sample diagram and physical images for the tensile shear test are shown in Figure 8a–c, the length and width are 90 and 20 mm, respectively. The experimental results are shown in Figure 8d. The maximum shear force is about 81 Mpa, which meets the enterprise's requirement that the combined strength of aluminum-steel composite plate is greater than 60 Mpa. The interface between the aluminum and dovetail channel explosive composite plate did not separate during the shear test, the fracture is located on one side of the aluminum layer.

4. Conclusions

The HCECB method is put forward to produce 6061 Al alloy-Q235a composite plates. The results of the SEM and EDX analyses show that the element diffusion appeared at the welding interface of 6061-Q235a, which was complex and dense with good bonding quality. The microhardness of the composite interface and its vicinity is in line with the microhardness requirements of composite materials. The interface of the composite plate is combined in the form of flatness and microwave, with close bonding, and almost no melt. The shear test results of the composite plate show that the shear force is greater than 80 MPa, which meets the requirements of the combined strength of Al-steel composite plate. Therefore, the welding composite plate showed better mechanical properties, which would be enough to meet the commercial requirement. In summary, the related selected metal materials and the size of the metal plates can be unlimited, and the groove of the base plate has diversity, which can meet the needs of different business situations.

Author Contributions: Conceptualization, Z.-w.S., H.-h.M., and N.L.; methodology, H.-h.M.; formal analysis, Y.-l.S. and N.L.; investigation, Y.-l.S., M.Y., and L.-q.W.; writing—original draft preparation, Y.-l.S., M.Y., L.-q.W., and N.L.; writing—review and editing, Y.-l.S. and N.L.; supervision, H.-h.M. and N.L.; project administration, Z.-w.S.; funding acquisition, H.-h.M. and N.L. All authors have read and agreed to the published version of the manuscript.

Funding: This research was financially supported by the National Natural Science Foundation of China, Grant Nos.51874267, 120702363 and 51674229.

References

1. Çelikyürek, İ.; Torun, O.; Baksan, B. Microstructure and strength of friction-welded Fe–28Al and 316 L stainless steel. *Mater. Sci. Eng. A* **2011**, *528*, 8530–8536. [CrossRef]
2. Fan, M.; Domblesky, J.; Jin, K.; Qin, L.; Cui, S.; Guo, X.; Kim, N.; Tao, J. Effect of original layer thicknesses on the interface bonding and mechanical properties of TiAl laminate composites. *Mater. Des.* **2016**, *99*, 535–542. [CrossRef]
3. HSFL Carvalho, G.; Galvão, I.; Mendes, R.; M Leal, R.; Loureiro, A. Aluminum-to-Steel Cladding by Explosive Welding. *Metals* **2020**, *10*, 1062. [CrossRef]

4. Upadhyay, A.; Sherpa, B.B.; Kumar, S.; Srivastav, N.; Dinesh Kumar, P.; Agarwal, A. Experimental Investigation and Micro-Structure Study of Interface of Explosive Welded SS304 and AA6061 Plates. *Mater. Sci. Forum* **2015**, *830*, 261–264. [CrossRef]

5. Elango, E.; Saravanan, S.; Raghukandan, K. Experimental and numerical studies on aluminum-stainless steel explosive cladding. *J. Cent. South Univ.* **2020**, *27*, 1742–1753. [CrossRef]

6. Han, J.H.; Ahn, J.P.; Shin, M.C. Effect of interlayer thickness on shear deformation behavior of AA5083 aluminum alloy/SS41 steel plates manufactured by explosive welding. *J. Mater. Sci.* **2003**, *38*, 13–18. [CrossRef]

7. Zhang, Y.; Long, B.; Meng, K.; Gohkman, A.; Cuia, Y.; Zhang, Z. Diffusion bonding of Q345 steel to zirconium using an aluminum interlayer. *J. Mater. Process. Technol.* **2020**, *275*, 116352. [CrossRef]

8. Li, Y.; Liu, Y.; Yang, J. First principle calculations and mechanical properties of the intermetallic compounds in a laser welded steel/aluminum joint. *Opt. Laser Technol.* **2020**, *122*, 105875. [CrossRef]

9. Tanaka, T.; Nezu, M.; Uchida, S.; Hirata, T. Mechanism of intermetallic compound formation during the dissimilar friction stir welding of aluminum and steel. *J. Mater. Sci.* **2020**, *55*, 3064–3072. [CrossRef]

10. Li, L.; Xia, H.; Tan, C.; Ma, N. Influence of laser power on interfacial microstructure and mechanical properties of laser welded-brazed Al/steel dissimilar butted joint. *J. Manuf. Process.* **2018**, *32*, 160–174. [CrossRef]

11. Wang, X.; Li, F.; Huang, T.; Wang, X.; Liu, H. Experimental and numerical study on the laser shock welding of aluminum to stainless steel. *Opt. Lasers Eng.* **2019**, *115*, 74–85. [CrossRef]

12. Zhang, Y.; Huang, J.; Cheng, Z.; Ye, Z.; Chi, H.; Peng, L.; Chen, S. Study on MIG-TIG double-sided arc welding-brazing of aluminum and stainless steel. *Mater. Lett.* **2016**, *172*, 146–148. [CrossRef]

13. Babu, S.; Panigrahi, S.K.; Ram, G.J.; Venkitakrishnan, P.V.; Kumar, R.S. Cold metal transfer welding of aluminium alloy AA 2219 to austenitic stainless steel AISI 321. *J. Mater. Process. Technol.* **2019**, *266*, 155–164. [CrossRef]

14. Matsuda, T.; Adachi, H.; Sano, T.; Yoshida, R.; Hori, H.; Ono, S.; Hirose, A. High-frequency linear friction welding of aluminum alloys to stainless steel. *J. Mater. Process. Technol.* **2019**, *269*, 45–51. [CrossRef]

15. Murugan, B.; Thirunavukarasu, G.; Kundu, S.; Kailas, S.V. Influence of Tool Traverse Speed on Structure, Mechanical Properties, Fracture Behavior, and Weld Corrosion of Friction Stir Welded Joints of Aluminum and Stainless Steel. *Adv. Eng. Mater.* **2019**, *21*, 1800869. [CrossRef]

16. Thomä, M.; Gester, A.; Wagner, G.; Fritzsche, M. Analysis of the Oscillation Behavior of Hybrid Aluminum/Steel Joints Realized by Ultrasound Enhanced Friction Stir Welding. *Metals* **2020**, *10*, 1079. [CrossRef]

17. Chitturi, V.; Pedapati, S.R.; Awang, M. Challenges in dissimilar friction stir welding of aluminum 5052 and 304 stainless steel alloys. *Materialwissenschaft und Werkstofftechnik* **2020**, *51*, 811–816. [CrossRef]

18. Li, P.; Chen, S.; Dong, H.; Ji, H.; Li, Y.; Guo, X.; Yang, G.; Zhang, X.; Han, X. Interfacial microstructure and mechanical properties of dissimilar aluminum/steel joint fabricated via refilled friction stir spot welding. *J. Manuf. Process.* **2020**, *49*, 385–396. [CrossRef]

19. Kapil, A.; Mao, Y.; Vivek, A.; Cooper, R.; Hetrick, E.; Daehn, G. A new approach for dissimilar aluminum-steel impact spot welding using vaporizing foil actuators. *J. Manuf. Process.* **2020**, *58*, 279–288. [CrossRef]

20. Wang, J.; Lu, X.; Cheng, C.; Li, B.; Ma, Z. Improve the quality of 1060Al/Q235 explosive composite plate by friction stir processing. *J. Mater. Res. Technol.* **2020**, *9*, 42–51. [CrossRef]

21. Singh, P.; Deepak, D.; Brar, G.S. Comparative evaluation of aluminum and stainless steel dissimilar welded joints. *Mater. Today.* in press. [CrossRef]

22. Hokamoto, K.; Chiba, A.; Fujita, M.; Izuma, T. Single-shot explosive welding technique for the fabrication of multilayered metal base composites: Effect of welding parameters leading to optimum bonding condition. *Compos. Eng.* **1995**, *5*, 1069–1079. [CrossRef]

23. Aceves, S.M.; Espinosa-Loza, F.; Elmer, J.W.; Huber, R. Comparison of Cu, Ti and Ta interlayer explosively fabricated aluminum to stainless steel transition joints for cryogenic pressurized hydrogen storage. *Int. J. Hydrog. Energy* **2015**, *40*, 1490–1503. [CrossRef]

24. Li, X.; Ma, H.; Shen, Z. Research on explosive welding of aluminum alloy to steel with dovetail grooves. *Mater. Des.* **2015**, *87*, 815–824. [CrossRef]

25. Lysak, V.I.; Kuzmin, S.V. Energy balance during explosive welding. *J. Mater. Process. Technol.* **2015**, *222*, 356–364. [CrossRef]

26. Ming, Y.A.; Shen, Z.W.; Chen, D.G.; Deng, Y.X. Microstructure and mechanical properties of Al-Fe meshing bonding interfaces manufactured by explosive welding. *Trans. Nonferr. Met. Soc. China* **2019**, *29*, 680–691.

27. *GB/T 6386-2008 Test Method Clad Steel Plates-Mechanical and Technological Test*; China National Standardization Management Committee: Beijing, China, 2008.

28. Fan, Z.; Yu, H.; Li, C. Interface and grain-boundary amorphization in the Al/Fe bimetallic system during pulsed-magnetic-driven impact. *Scr. Mater.* **2016**, *110*, 14–18. [CrossRef]

Publisher's Note: MDPI stays neutral with regard to jurisdictional claims in published maps and institutional affiliations.

Article

Effect of Cr Atom Plasma Emission Intensity on the Characteristics of Cr-DLC Films Deposited by Pulsed-DC Magnetron Sputtering

Guang Li [1,2,†], **Yi Xu** [1,2,†] **and Yuan Xia** [1,2,*]

1   Institute of Mechanics, Chinese Academy of Sciences, Beijing 100190, China; lghit@imech.ac.cn (G.L.); xuyi@imech.ac.cn (Y.X.)

2   Center of Materials Science and Optoelectronics Engineering, University of Chinese Academy of Sciences, Beijing 100049, China

*   Correspondence: xia@imech.ac.cn

†   These authors contributed equally to this work.

Received: 4 June 2020; Accepted: 27 June 2020; Published: 28 June 2020

Abstract: A pulsed-dc (direct current) magnetron sputtering with a plasma emission monitor (PEM) system was applied to synthesize Cr-containing hydrogenated amorphous diamond-like carbon (Cr-DLC) films using a large-size industrial Cr target. The plasma emission intensity of a Cr atom at 358 nm wavelength was characterized by optical emission spectrometer (OES). C_2H_2 gas flow rate was precisely adjusted to obtain a stable plasma emission intensity. The relationships between Cr atom plasma emission intensity and the element concentration, cross-sectional morphology, deposition rate, microstructure, mechanical properties, and tribological properties of Cr-DLC films were investigated. Scanning electron microscope and Raman spectra were employed to analyze the chemical composition and microstructure, respectively. The mechanical and tribological behaviors were characterized and analyzed by using the nano-indentation, scratch test instrument, and ball-on-disk reciprocating friction/wear tester. The results indicate that the PEM system was successfully used in magnetron sputtering for a more stable Cr-DLC deposition process.

Keywords: diamond-like carbon (DLC); plasma emission intensity; magnetron sputtering; tribological performance

---

1. Introduction

Metal-containing hydrogenated amorphous diamond-like carbon films (Me-DLC) have raised much attention for decades because of their promising performance, such as high nano-hardness, chemical inertness, excellent tribological properties, and thermal stability [1–4]. By the doping of metals, a two-dimensional array of nanoclusters within the DLC matrix or an atomic-scale composite can be formed, and these nanostructures are responsible for the desirable mechanical and thermal properties [5–7]. Generally, Me-DLC films are synthesized by hybrid magnetron sputtering using a metal target with a hydrocarbon gas [8,9]. However, an inherent feature in this conventional hybrid deposition process, combining magnetron and plasma assisted chemical vapor deposition, is high instability due to the complex relation between the fluxes of reactive gas and sputtered metal from the target. During the deposition process, it inevitably suffers from the "hysteresis effect", immediate rapid transition from metal to poisoned state [10,11]. Such behavior originates from the formation of carbide on the target surface, leading to a significant decrease in the sputtering yield or a non-reproducible quality of films. Therefore, precise control of the 'poisoned' state of the metal target surface is vital for the deposition of high quality Me-DLC films.

Plasma emission monitoring (PEM) system, based on the gas flow control technique, has received extensive concern as an important method of stabilizing the deposition process in the "transition" state because of its high accuracy and stability [12,13]. Wu et al. showed that various CrN_X films were obtained under a stable deposition mode by adjusting the N_2 gas flow rate and exhibiting different Cr atom plasma emission intensities [14]. Ohno et al. stated that stable depositions for TiO_2 films were achieved in the transition region with the PEM system [15]. Hirohata et al. reported the sputtering of ZnAl in O_2 atmosphere with the PEM system and showed that ZnO films deposition rates were about 15 times higher than those by conventional magnetron sputter [16]. In a similar way, Yang et al. investigated the properties of Ti-containing and W-containing hydrogenated DLC films deposited via high power impulse magnetron sputtering (HiPIMS), by controlling the C_2H_2 flow rate and different Ti^{2+} plasma emission intensities under the PEM system [17,18]. They found that the hysteresis curve depicted almost completely overlapping curves with some fluctuations in the amount of C_2H_2, which implied that the PEM controlled HiPIMS process was easy, accurate, and stable, especially in reactive poisoning deposition. As can be clearly seen from these publications, in comparison to conventional hybrid magnetron sputtering, the PEM system exhibits excellent capabilities for improving film deposition rate and stabilizing the deposition process. On the other hand, however, the sputtered targets in most of these works are small in size. For example, 200 mm-diameter Fe and Zr targets were chosen in the publications [12,13], respectively. Only a few of them concern the deposition of films using a large-sized target under the PEM system. Compared with the small sized targets, the large sized targets are more suitable for industrial production. Due to the larger sputtered area, the target poisoning and abnormal discharge (arcing) phenomenon for large sized targets were more serious, which resulted in an unstable process or a nonreproducible quality of thin films for industrial application.

In this work, a pulsed-dc magnetron sputtering with PEM system was applied to fabricate Cr-DLC films using a large-size industrial Cr target, and C_2H_2 as feeding gas. The synthesis and characteristics were examined to evaluate the feasibility of a PEM system. To this end, the stability of Cr-DLC film deposition process with the PEM system was compared to that of conventional hybrid magnetron sputtering. The effect of Cr atom plasma emission intensity on the element concentration, cross-sectional morphology, deposition rate, microstructure, mechanical properties, and tribological properties of Cr-DLC films were explored. Our results demonstrate that the PEM system is an effective method for stabilizing the deposition process of the Cr-DLC films at different Cr atom plasma emission intensities, and the Cr-DLC film deposition rate is as high as 92.7 nm/min.

2. Experiment Details

2.1. Deposition

As shown in Figure 1, the experiments were conducted with an unbalanced magnetron sputtering system. A closed-loop plasma emission monitor system (PEM, Nava Fabrica FlotronTM, Vilnius, Lithuania) was utilized to precisely control the C_2H_2 gas flow rate. The plasma emission intensity of the Cr atom at 358 nm (see Figure 2) was characterized by a collimator of optical emission spectrometer (OES, Fabrica Flotron TM, Vilnius, Lithuania), which was placed 50 mm away from the target. During the deposition process, a pulsed DC power (MSP-20D, Pulse Tech, Nanjing, China) was used to power the Cr target, while a DC power (MSP-50D, Pulse Tech, Nanjing, China) was applied to the substrate. In this work, the set values of 100% and 0% were defined as the pure metallic deposition and fully poisoning deposition, respectively. The set values between 0% and 100% represent the percentage of the reduced Cr atom plasma emission intensity.

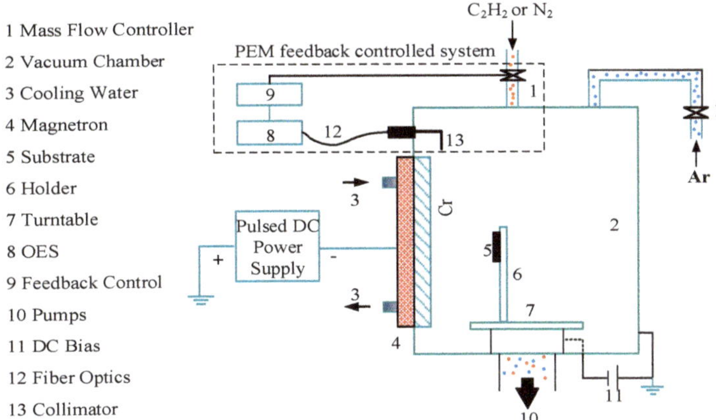

Figure 1. Schematics of deposition apparatus for Cr-containing hydrogenated amorphous diamond-like carbon (Cr-DLC) films.

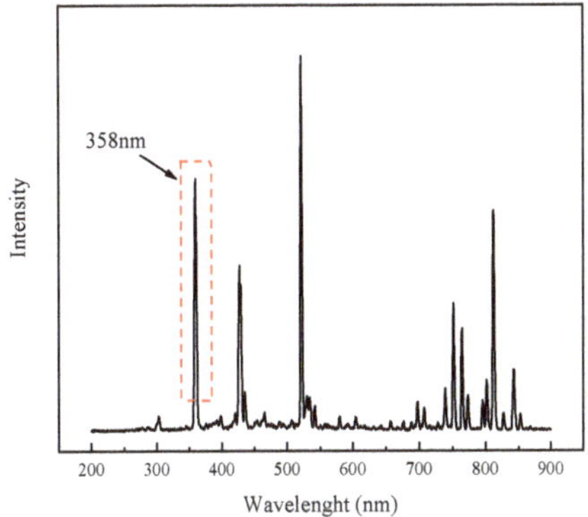

Figure 2. Typical optical emission spectrometer (OES) spectra during sputtering Cr target powered by pulsed-dc (direct current) magnetron sputtering.

There were two kinds of samples mounted in the vacuum chamber: high speed steel (HSS) samples (ϕ 30 mm) and silicon wafer (5 × 5 mm^2). The vacuum chamber was mounted vertically with rectangular Cr (99.99% purity) target (131 × 580 mm^2). The base vacuum of the chamber was less than 5 × 10^{-3} Pa. After being ultrasonically pre-cleaned, all samples were further cleaned for 20 min in an Ar plasma environment (120 sccm, 2 Pa, −900 V bias voltage) in the chamber to remove surface contaminants. Then, the deposition of the layered interlayer containing Cr (150 nm) and CrN (500 nm) were deposited at a bias of −75 V in sequence at an Ar gas deposition pressure of 0.5 Pa in a pulsed DC mode to power the target to 5 A with a frequency of 40 kHz and a duty ratio of 80%. The CrN films were deposited under the PEM system, and the set value was fixed at 60%. The Cr/CrN interlayers were pre-deposited in order to improve the adhesion of the Cr-DLC films to the substrates. The deposition parameters for the Cr-DLC films are listed in detail in Table 1. During the deposition process, the substrate holder is stationary, and no additional heading was applied. For all deposition conditions, the substrate temperature was below 75 °C.

Table 1. The deposition parameters of Cr-DLC films.

Parameter	Value
Target-Substrate Distance (mm)	85
Ar Gas Pressure (Pa)	0.5
Bias Voltage (V)	−75
Set Value (%)	2/5/10/15/20
Pulse Frequency (kHz)	40
Duty Ratio (%)	80
Pulse Current (A)	5
Average power (kW)	1.44/1.42/1.38/1.36/1.33 (stable deposition state)
Deposition Time (min)	50

2.2. Characterization

A scanning electron microscope (FE-SEM, XL30 S-FEG, Philips, Amsterdam, Netherlands) with energy dispersive spectrometer (EDS, Philips, Amsterdam, Netherlands) was used to obtain the chemical composition and cross-sectional morphology. Raman spectra were analyzed by using an excitation wavelength of 523 nm and an argon ion laser, to characterize the chemical substructures of Cr-DLC films. The nano-hardness and modulus were investigated by a nano-indentation tester (Nano-Indentor G200, Agilent, SC, USA) with a load precision of 50 nN in continuous stiffness method, and an indentation of 300 nm was selected to avoid the substrate effect. A scratch test instrument (MFT-R4000, Huahui, Lanzhou, China) was employed to study the adhesion strength. The load was gradually increased from 0 to 50 N with 5 mm/min at a loading speed of 100 N/min. The friction coefficients and wear rates were evaluated by a ball-on-disk reciprocating friction/wear tester (MFT-R4000, Huahui, Lanzhou, China) at a relative humidity of approximately 22% RH and temperature of 23 °C. GCr15 bearing steel balls 5mm in diameter were used against the films at 15 N (F) and a sliding speed of 240 mm/min (V). The track length (D) and the duration time (T) were 5 mm and 60 min, respectively. After the wear experiments, the profiles of the tracks were measured by a surface profiler (MFT-R4000, Huahui, Lanzhou, China), and the cross-sectional areas of the wear tracks (A) were measured on each wear track at three points. The wear rates of the Cr-DLC films were calculated as follows: $(A \times D)/(F \times V \times T)$.

3. Results and Discussion

3.1. Stability Comparison: Constant Flow Control Mode vs. PEM System

A typical set value-time relationship for Cr-DLC depositing under constant flow control mode (53 sccm) is plotted in Figure 3a. The set value of Cr is about 12.4% in the initial stage. With the increase in time, the set value of Cr dramatically decreases, and then gradually approaches about 6% at 1100 s. For times greater than 1100 s, the set value of Cr kept a relatively stable value. The set value of Cr at 1100 s is nearly two times lower than that at the beginning. These phenomena can be attributed to the formation of carbide on the target surface, which reduces the number of sputtered particles from the Cr targets, thereby leading to the "target poisoning" effect [19,20]. Therefore, it is obvious that the constant flow control mode does not have the capacity for offering a stable deposition process. Here, PEM system offers a precise control and stable operation of the deposition process in Figure 3b. As the deposition of Cr-DLC films starts, the set value of Cr decreases quickly to the set value and then stays stable. By automatically decreasing the C_2H_2 gas flow rate from 61 to 53 sccm, the set value remains a stable state throughout the following deposition process. This dynamic process can be understood as follows: sputtering of the carbide on the Cr target is increased by an automatic decrease of the C_2H_2 flow rate in the PEM system. Subsequently, a stable Cr emission intensity can be maintained, which results in the Cr target surface being stabilized in the "transition" state, rather than 'poisoned' state. The target voltages and pressures during the deposition process under the PEM system at different set values are shown in Figure 4a,b, respectively. Like the set value of Cr above, the target voltage and the deposition pressure both remain stable and no obvious drift occurred to either side during the deposition process, except for a very short period of adjustment at the beginning. Meanwhile, as the pulse current is kept constant (5A), the average power has the same variation tendency to the target voltage, and the average powers at the stable deposition state are 1.44 kW at 2%, 1.42 kW at 5%, 1.38 kW at 10%, 1.36 kW at 15%, and 1.33 kW at 20%. These results further indicate that the PEM system is an effective method for stabilizing the deposition process of the Cr-DLC films.

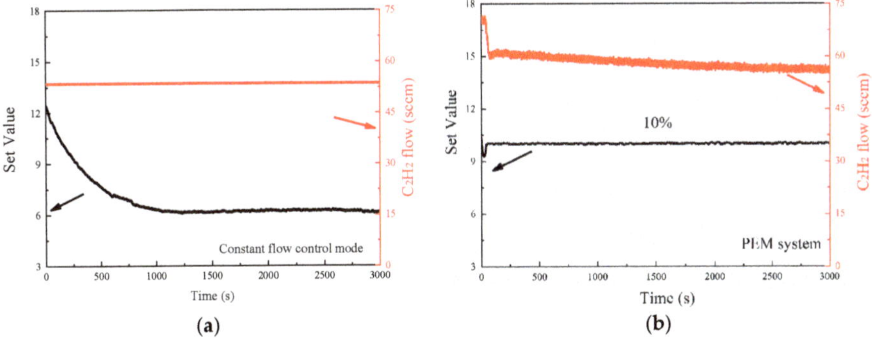

Figure 3. Experimental time-dependent emission intensity of Cr atom and C_2H_2 flow rate under (**a**) constant flow control mode and (**b**) plasma emission monitoring (PEM) system at a set value of 10%. The 0 s is the time of setting the C_2H_2 gas flow rate for constant flow control and set value for the PEM system.

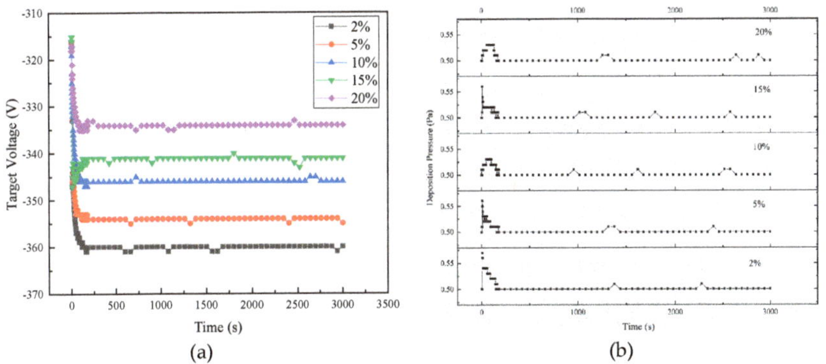

Figure 4. Experimental time-dependent (**a**) target voltages and (**b**) deposition pressures under the PEM system at different set values.

3.2. Effect of Set Values on the Composition and Deposition Rate

Table 2 lists the chemical composition of the Cr-DLC films deposited at different set values. Due to the fact that the hydrogen could not be detected by EDS, the relative atomic composition ratio was considered a sum of chromium, carbon, and oxygen. The relative content of C decreased from 97.59 to 82.34 at.% as the set value increased from 2% to 20%, which was likely due to the lower amount of C_2H_2 gas that was introduced. The Cr relative content increased from 1.96 to 17.21 at.%, with a decreased degree of target poisoning. In addition, there was a small amount of oxygen in the films, which may be caused by the contamination from the residual vacuum.

Table 2. Composition of Cr-DLC films.

Set Value	Element Composition (at.%)		
	C	Cr	O
2%	97.59	1.96	0.45
5%	93.88	5.78	0.34
10%	89.87	9.26	0.87
15%	86.11	13.61	0.28
20%	82.34	17.21	0.45

The cross-sectional morphologies of Cr-DLC films deposited at different set values are illustrated in Figure 5. It can be seen that the films all exhibit a typical dense morphology. According to the images in Figure 5, it was observed that the film thicknesses at the set values of 2%, 5%, 10%, 15%, and 20% were 4.6, 3.8, 3.6, 3.3 and 3.1 μm, respectively. The deposition rates for Cr-DLC films were obtained by dividing the thickness with the deposition time, and further exhibited in Figure 6. The film deposition rate exhibits a negative relationship with the set value. When the set value was 2%, the Cr-DLC film deposition rate was as high as 92.7 nm/min. Although the deposition rate decreased to 62.9 μm/h at a set value of 20%, the deposition of these Cr-DLC films under the PEM system presented a higher film deposition rate than that of conventional hybrid magnetron sputter deposition processes [9,21–24].

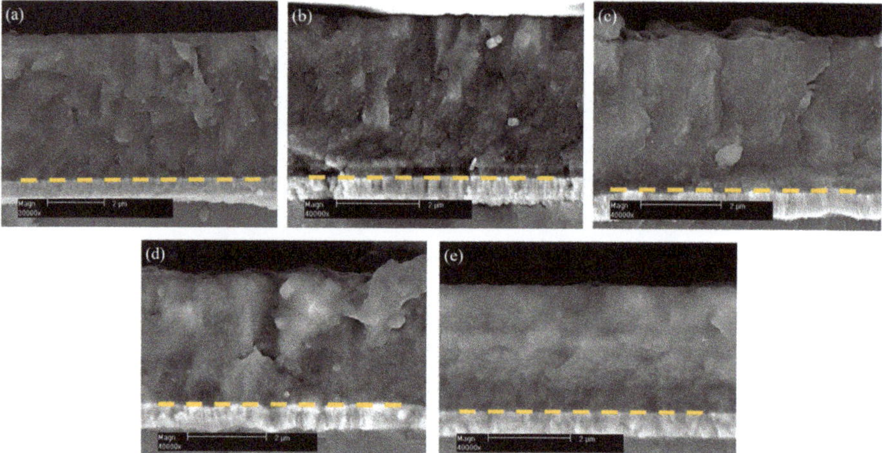

Figure 5. Cross-sectional morphologies of Cr-DLC films deposited at set values of: (**a**) 2%, (**b**) 5%, (**c**) 10%, (**d**) 15%, and (**e**) 20%. The start of the Cr-DLC films was marked by a yellow line in each figure.

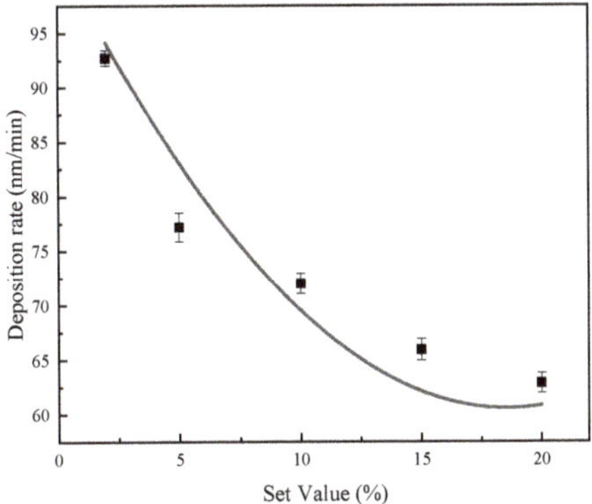

Figure 6. Deposition rates of Cr-DLC films as a function of the set value.

3.3. Effect of Set Values on the Microstructure

Raman spectra is one of the popular and effective nondestructive tools used to differentiate and acquire the bonding nature of DLC films. Figure 7 displays the Raman spectra between 1000 and 1800 cm^{-1} for Cr-DLC films. In general, the Raman spectra can be fitted by two Gaussian distributions: one centered at about 1350 cm^{-1}, corresponding to the D peak of the disordered structure; and the another one centered at about 1580 cm^{-1}, related to the G peak of the graphite structure [25–27]. The integrated intensity ratio of the D and G band, I_D/I_G, can be correlated with the C–C sp^3/sp^2 bonding ratio [28]. Empirically, it is widely accepted that a larger I_D/I_G ratio and higher G peak position means a lower C–C sp^3 content for non-hydrogenated DLC films [29,30]. The G peak positions slightly shift to a higher Raman frequency with an increasing set value. In Figure 8, the I_D/I_G ratio increases from 1.94 to 2.76 when the set value increases from 2% to 20%, respectively. Nevertheless, for hydrogenated DLC films, this ratio influenced both C–C and C–H sp^3 bonding. Combined with the previous reports for the hydrogenated Cr-DLC [23,31,32], it was estimated that the sp^3/sp^2 ratio was approximately 0.4 for Cr-DLC films at the low set value, and this ratio decreases with the increase in set value. Such variations in the sp^3/sp^2 ratio with set value can be explained by the fact that a lower amount of C_2H_2 gas is introduced at a higher set value. As a result, more Cr and less hydrogen are incorporated into the Cr-DLC films. On the other hand, the overall Raman intensity of sp^2 bonds decrease with the increase of the set value, which means that the proportion of the DLC phase decreases.

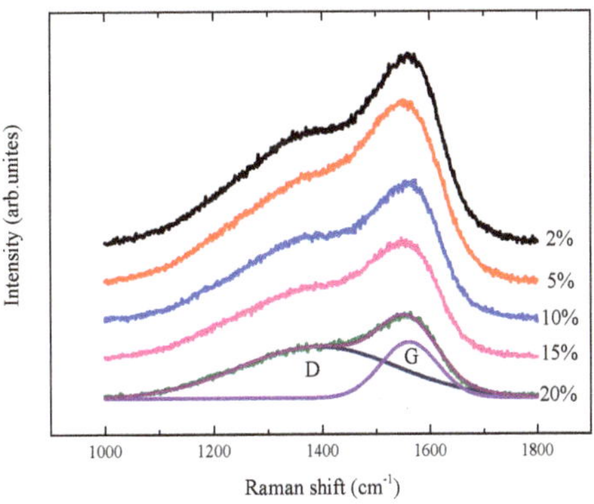

Figure 7. Raman spectra of the Cr-DLC films deposited at different set values.

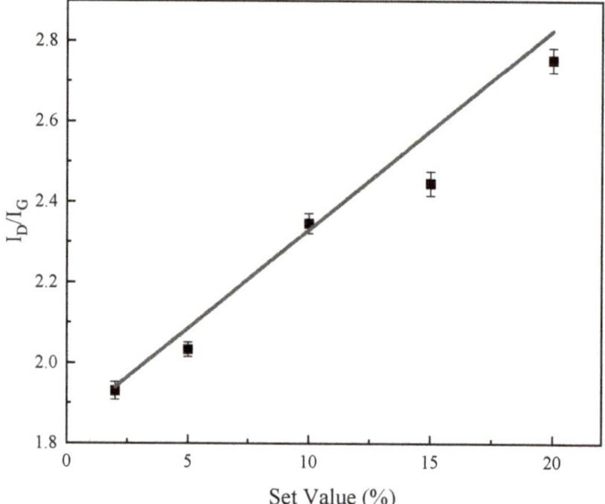

Figure 8. I_D/I_G ratios of Cr-DLC films deposited at different set values.

3.4. Effect of Set Values on Mechanical Properties

The hardness and elastic modulus of the Cr-DLC films deposited at different set values are depicted in Figure 9. The hardness and elastic modulus increased monotonously from 5.72 to 8.02 GPa and from 80.19 to 139.63 GPa, respectively. According to the above Raman results, the proportion of the DLC phase tends to decrease when the set value increases, resulting in a decrease of hardness. However, with the increasing set value, more Cr atoms are incorporated into the Cr-DLC films, and hard chromium carbide nano-particles are formed and embedded in the DLC matrix, which significantly contributes to an increase in hardness and elastic modulus [8]. Based on the above analysis, we can conclude that hard chromium carbide phase was formed to offset the effect of the decrease of DLC phase that contributed to decreasing hardness and elastic modulus when set value increased from 2% to 20%. Overall, the hardness of the Cr-DLC films were quite low. Previous reports described similar phenomena for Cr-containing hydrogenated amorphous DLC films [33–35]. This is considered to be related to the higher hydrogen content in films. Excess hydrogen is bonded to carbon and leads to more C–H bonding in the Cr-DLC films and reduces the amount of C–C bonds forming the amorphous network, which reduces the strength of the amorphous phase. Furthermore, the lower deposition temperature of 75 °C is also another vital reason behind the relatively low hardness for Cr-DLC films.

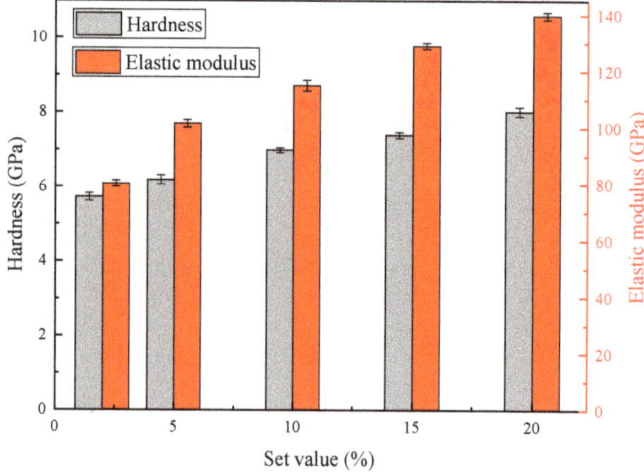

Figure 9. Hardness and Elastic modulus of the Cr-DLC films deposited at different set values.

Obtaining the high adhesion strength is one of the major technological challenges for the DLC films, and it plays an important role in the wear and corrosion resistance of the films. Figure 10 shows the critical loads (LC_2) of the Cr-DLC films deposited at different set values, and their critical loads are 15, 22, 18, 15 and 12N, respectively. The LC_2 is associated with the start of chipping failure extending from the arc tensile cracks, indicating adhesive failure between the coating and the substrate. The results indicate that the adhesion initially increased and then decreased when set value >5%. Generally speaking, the adhesion strength of films mainly depends on the film internal stress. In the initial stage, increasing set value meant more Cr content in the Cr-DLC film, resulting in a decrease in the proportion of DLC phase (confirmed by Raman spectra, see Figure 8), which was beneficial to the release of internal stress [36,37]. Therefore, the films possess a relatively lower internal stress when the set value increases up to 5%, which may improve the adhesion strength. Moreover, Choi et al. pointed out that metal doping reduced the directionality of bond, which led to a decrease in internal stress caused by bond angle distortion in the amorphous carbon network [38]. However, plenty of carbide phases will be generated with further increases of the set value. The Cr–C bond length is longer than the C–C bond length, causing an increase in internal stress [39,40]. As a result, the adhesion strength decreased as set value increased from 5% to 20%.

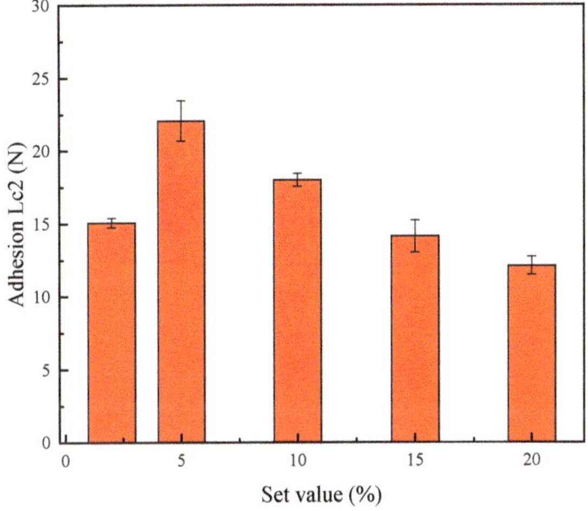

Figure 10. Adhesion of the Cr-DLC films deposited at different set values.

3.5. Effect of Set Values on Tribological Performance

Figure 11 depicts the friction coefficients evolution of the Cr-DLC films deposited at different set values, with the sliding time and corresponding surface profiles of the tracks after test. A smooth and indistinct track was observed for the films deposited at the set value of 2%, while the Cr-DLC films deposited at the set value > 5% presented deeper and broader wear tracks. Figure 12 shows the wear surfaces on the Cr-DLC films deposited at the set value of 2% and 20%. The wear track of the films deposited at a set value of 2% are larger than those deposited at a set value of 20%, which implied that the higher set value (meaning higher Cr content) deteriorated tribological performance of Cr-DLC films. Figure 13 shows the average friction coefficients, calculated after about 10 min of the sliding time, and the wear rates. The average friction coefficient and wear rate exhibited a relatively low value of 0.14 and 1.21×10^{-9} mm^3/Nm at the set value of 2%, respectively. With the increase of set value, the average friction coefficient and wear rate both presented an increased tendency. The average friction coefficient and wear rate with set values of 5%, 10%, 15%, and 20% were 0.17 and 1.94×10^{-9} mm^3/Nm, 0.18 and 2.16×10^{-9} mm^3/Nm, 0.21 and 2.88×10^{-9} mm^3/Nm, and 0.22 and 5.76×10^{-9} mm^3/Nm, respectively. For the films deposited at low set values (low Cr content), the films displayed the characteristics of an amorphous carbon structure, which led to the lower friction coefficient and wear rate. Nevertheless, higher Cr content led to the formation of a hard chromium carbide phase, and the amorphous Cr-DLC film was gradually transformed into carbide-rich film [41–43]. Thus, sliding between hard materials induced high shear strength, which in turn yielded high friction coefficients. These processes can yield abrasive wear, causing the deterioration of the tribological properties. On the other hand, the friction coefficients and wear properties were dominated by Cr metal at higher Cr content. During the sliding process, Cr metal interacted with the oxygen in the atmosphere, resulting in the formation of metal oxide [44,45], which increased the friction coefficient and the wear rate.

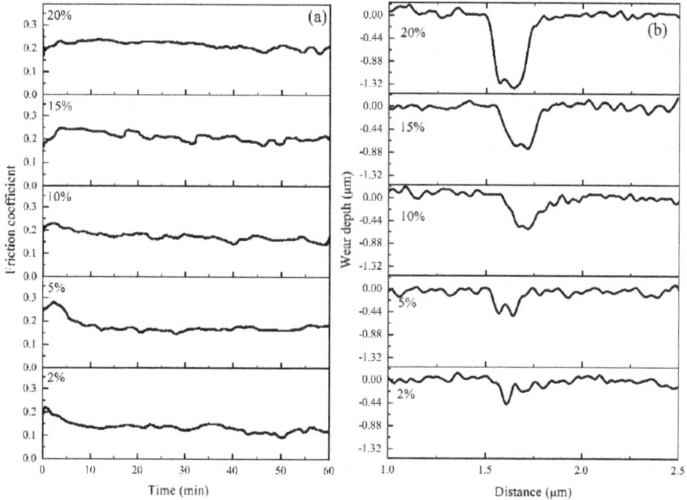

Figure 11. (a) Friction coefficient of the Cr-DLC films deposited at different set values and (b) corresponding surface profiles of the wear tracks after test.

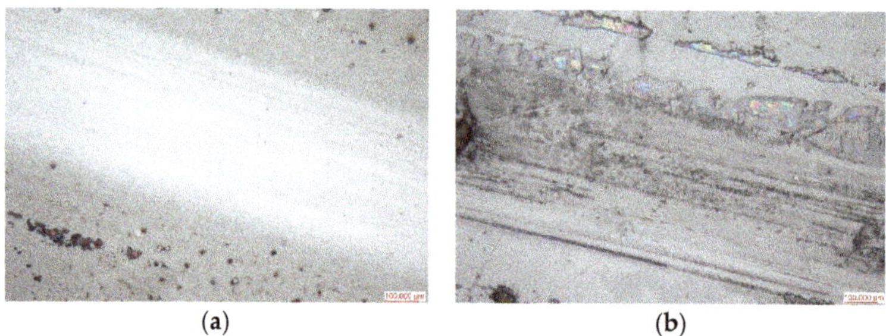

(a)   (b)

Figure 12. Wear tracks micrographs obtained by optical microscope of the Cr-DLC films deposited at set values of 2% (a) and 20% (b).

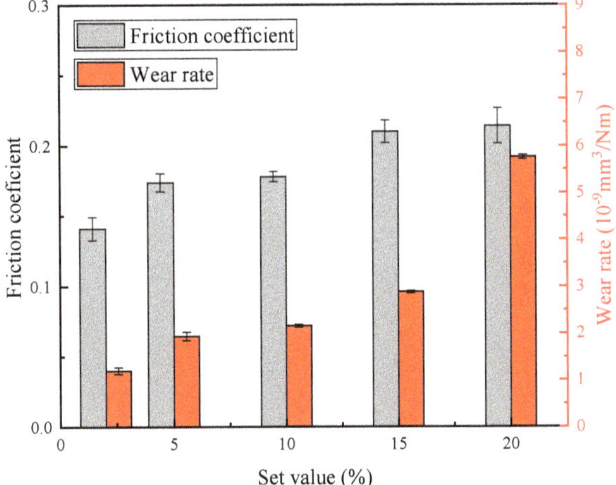

Figure 13. The average friction coefficient and wear rate of the Cr-DLC films deposited at different set values.

4. Conclusions

In summary, Cr-DLC films were synthesized by pulsed-dc magnetron sputter with a PEM system using a large-size industrial Cr target. The relationships between the Cr atom plasma emission intensity (set value), target voltage, pressure, and time suggest that the PEM system is an effective method for stabilizing the deposition process in the "transition" state. Compositional analysis showed that the Cr content increased from 1.96 to 17.21 at.%, and the set value increased from 2% to 20%. The Cr-DLC film deposition rate was as high as 92.7 nm/min at the set value of 2%, which is obviously higher than that for conventional hybrid magnetron sputtering. The adhesion strength increased initially and then reached its maximum values of 22 N at the set value of 5%. Raman results demonstrated that increasing set values promotes the graphitizing of Cr-DLC films. The hardness was enhanced with the increase of set value. In addition, it was observed that both the friction coefficient and wear rate increased with set value, and the lowest friction coefficient and wear rate were 0.14 and 1.2×10^{-9} mm^3/Nm, respectively. Overall, the hardness values reported in this work were quite small; more experiments and deeper analysis are needed to reveal the mechanism behind it. On the other hand, there is a growing demand for over-sized targets (Length > 800 mm) in industrial applications. Due to the larger sputtered area, a single collimator is not enough to monitor the plasma emission intensity for the entire area. We are now designing multi-collimators and corresponding installation locations. More work is being conducted in this direction, and the results will be reported in due course.

Author Contributions: Conceptualization, methodology, supervision, G.L.; formal analysis, investigation, data curation, writing—original draft preparation, writing—review and editing, Y.X. (Yi Xu); project administration, funding acquisition, Y.X. (Yuan Xia). All authors have read and agreed to the published version of the manuscript.

Funding: This research was jointly found by National Nature Foundation of China (No. 51871230 & 51701229) and the Strategic Priority Research Program of the Chinese Academy of Sciences (No. XDB22040503).

Conflicts of Interest: The authors declare no conflict of interest.

References

1. Sun, L.; Zuo, X.; Guo, P.; Li, X.; Ke, P.; Wang, A. Role of deposition temperature on the mechanical and tribological properties of Cu and Cr co-doped diamond-like carbon films. *Thin Solid Films* **2019**, *678*, 16–25. [CrossRef]
2. Zhou, Y.; Guo, P.; Sun, L.L.; Liu, L.L.; Xu, X.W.; Li, W.X.; Lee, K.R.; Wang, A.Y. Microstructure and property evolution of diamond-like carbon films co-doped by Al and Ti with different ratios. *Surf. Coat. Technol.* **2019**, *361*, 83–90. [CrossRef]
3. Savchenko, D.; Vorlíček, V.; Prokhorov, A.; Kalabukhova, E.; Lančok, J.; Jelínek, M. Raman and EPR spectroscopic studies of chromium-doped diamond-like carbon films. *Diam. Relat. Mater.* **2018**, *83*, 30–37. [CrossRef]
4. Sun, L.L.; Zuo, X.; Guo, P.; Li, X.W.; Ke, P.L.; Wang, A.Y. Structural properties of hydrogenated Al-doped diamond-like carbon films fabricated by a hybrid plasma system. *Thin solid Films* **2019**, *678*, 16–25. [CrossRef]
5. Zhou, Y.F.; Li, L.L.; Shao, W.; Chen, Z.H.; Wang, S.F.; Xing, X.L.; Yang, Q.X. Mechanical and tribological behaviors of Ti-DLC films deposited on 304 stainless steel: Exploration with Ti doping from micro to macro. *Diam. Relat. Mater.* **2020**, *107*, 107870. [CrossRef]
6. Singh, V.; Palshin, V.; Tittsworth, R.; Meletis, E. Local structure of composite Cr-containing diamond-like carbon thin films. *Carbon* **2006**, *44*, 1280–1286. [CrossRef]
7. Xu, X.W.; Zhou, Y.; Liu, L.L.; Guo, P.; Li, W.X.; Lee, K.R.; Cui, P.; Wang, A.Y. Corrosion behavior of diamond-like carbon film induced by Al/Ti co-doping. *Appl. Surf. Sci.* **2020**, *590*, 144877. [CrossRef]
8. Zhou, S.; Liu, L.; Ma, L.; Wang, Y.; Liu, Z. Influence of CH4 Flow Rate on Microstructure and Properties of Ti-C: H Films Deposited by DC Reactive Magnetron Sputtering. *Tribol. Trans.* **2017**, *60*, 852–860. [CrossRef]
9. Santiago, J.A.; Martine, I.F.; Lopez, J.C.S.; Rojas, T.C.; Wennberg, A.; Gonzalez, V.B.; Aldareguia, J.M.; Monclus, M.A.; Arrabal, R.G. Tribomechanical properites of hard Cr-doped DLC coatings deposited by low-frequency HiPIMS. *Surf. Coat. Technol.* **2020**, *382*, 124899. [CrossRef]
10. Shimizu, T.; Villamayor, M.; Lundin, D.; Helmersson, U. Process stabilization by peak current regulation in reactive high-power impulse magnetron sputtering of hafnium nitride. *J. Phys. D Appl. Phys.* **2016**, *49*, 065202. [CrossRef]
11. Audronis, M.; Bellido-Gonzalez, V.; Daniel, B. Control of reactive high power impulse magnetron sputtering processes. *Surf. Coat. Technol.* **2010**, *204*, 2159–2164. [CrossRef]
12. Aubry, E.; Liu, T.; Dekens, A.; Perry, F.; Mangin, S.; Hauet, T.; Billard, A. Synthesis of iron oxide films by reactive magnetron sputtering assisted by plasma emission monitoring. *Mater. Chem. Phys.* **2019**, *223*, 360–365. [CrossRef]
13. Coddet, P.; Pera, M.; Billard, A. Reactive co-sputter deposition of YSZ coatings using plasma emission monitoring. *Surf. Coat. Technol.* **2011**, *205*, 3987–3991. [CrossRef]
14. Wu, W.Y.; Hsiao, B.H.; Chen, P.-H.; Chen, W.-C.; Ho, C.-T.; Chang, C.-L. CrNx films prepared using feedback-controlled high power impulse magnetron sputter deposition. *J. Vac. Sci. Technol. A Vac. Surf. Films* **2014**, *32*, 02B115. [CrossRef]
15. Ohno, S.; Takasawa, N.; Sato, Y.; Yoshikawa, M.; Suzuki, K.; Frach, P.; Shigesato, Y. Photocatalytic TiO2 films deposited by reactive magnetron sputtering with unipolar pulsing and plasma emission control systems. *Thin Solid Films* **2006**, *496*, 126–130. [CrossRef]
16. Hirohata, K.; Nishi, Y.; Tsukamoto, N.; Oka, N.; Sato, Y.; Yamamoto, I.; Shigesato, Y. Al-doped ZnO (AZO) films deposited by reactive sputtering with unipolar-pulsing and plasma-emission control systems. *Thin Solid Films* **2010**, *518*, 2980–2983. [CrossRef]
17. Jiang, X.; Yang, F.-C.; Lee, J.-W.; Chang, C.-L. Effect of an optical emission spectrometer feedback-controlled method on the characterizations of nc-TiC/aC: H coated by high power impulse magnetron sputtering. *Diam. Relat. Mater.* **2017**, *73*, 19–24. [CrossRef]
18. Chang, C.L.; Yang, F.C. Synthesis and characteristics of nc-WC/a-C:H thin films deposited via a reactive HIPIMS process using optical emission spectrometry feedback control. *Surf. Coat. Technol.* **2018**, *350*, 1120–1127. [CrossRef]
19. Purandare, Y.P.; Ehiasarian, A.P.; Eh Hovsepian, P. Target poisoning during CrN deposition by mixed high power impulse magnetron sputtering and unbalanced magnetron sputtering technique. *J. Vac. Sci. Technol. A Vac. Surf. Films* **2016**, *34*, 041502. [CrossRef]

20. Layes, V.; Corbella, C.; Monjé, S.; der Gathen, V.S.; von Keudell, A.; de los Arcos, T. Connection between target poisoning and current waveforms in reactive high-power impulse magnetron sputtering of chromium. *Plasma Sour. Sci. Technol.* **2018**, *27*, 084004. [CrossRef]

21. Guo, T.; Kong, C.C.; Li, X.W.; Guo, P.; Wang, Z.Y.; Wang, A.Y. Microstructure and mechanical properties of Ti/Al co-doped DLC films: Dependence on sputtering current, source gas, and substrate bias. *Appl. Surf. Sci.* **2017**, *410*, 51–59. [CrossRef]

22. Huang, L.; Yuan, J.T.; Li, C. Influence of titanium concentration on mechanical properties and wear resistance to Ti6Al4V of Ti-C: H on cemented carbide. *Vacuum* **2017**, *138*, 1–7. [CrossRef]

23. Wu, Z.Z.; Tian, X.B.; Gui, G.; Gong, C.Z.; Yang, S.Q.; Chu, P.K. Microstructure and surface properties of chromium-doped diamond -like carbon thin films fabricated by high power pylsed magnetron sputtering. *Appl. Surf. Sci.* **2013**, *276*, 31–36. [CrossRef]

24. Tillmann, W.; Dias, N.F.; Stangier, D. Tribo-mechanical properties of CrC/a-C thin films sequentially deposited by HiPIMS and mfMS. *Surf. Coat. Technol.* **2018**, *335*, 173–180. [CrossRef]

25. Xu, Y.; Li, L.; Chu, P. Deposition of diamond-like carbon films on interior surface of long and slender quartz glass tube by enhanced glow discharge plasma immersion ion implantation. *Surf. Coat. Technol.* **2015**, *265*, 218–221. [CrossRef]

26. Chen, J.; Li, Y.; Huang, L.; Li, C.; Shi, G. High-yield preparation of graphene oxide from small graphite flakes via an improved Hummers method with a simple purification process. *Carbon* **2015**, *81*, 826–834. [CrossRef]

27. Duraia, E.-S.M.; Fahami, A.; Beall, G.W. Modifications of graphite and multiwall carbon nanotubes in the presence of urea. *J. Electron. Mater.* **2018**, *47*, 1176–1182. [CrossRef]

28. Beeman, D.; Silverman, J.; Lynds, R.; Anderson, M. Modeling studies of amorphous carbon. *Phys. Rev. B* **1984**, *30*, 870. [CrossRef]

29. Tucker, M.D.; Ganesan, R.; Mcculloch, D.G.; Partridge, J.G.; Stueber, M.; Ulrich, S.; Bilek, M.M.M.; Mckenzie, D.R.; Marks, N.A. Mixed-mode high-power impulse magnetron sputter deposition of tetrahedral amorphous carbon with pulse-length control of ionization. *J. Appl. Phys.* **2016**, *119*, 155303. [CrossRef]

30. Mbiombi, W.; Mathe, B.; Wamwangi, D.; Erasmus, R.; Every, A.; Billing, D.G. Optoelectronic and mechanical properties of PVD diamond-like carbon films. *Mater. Today Proc.* **2018**, *5*, 27307–27315. [CrossRef]

31. Bin, L.Q.; Zhang, B.; Zhou, Y.; Zhang, J.Y. Improving the internal stress and wear resistance of DLC film by low content Ti doping. *Solid State Sci.* **2013**, *20*, 17–22.

32. Yang, W.; Guo, Y.; Xu, D.; Li, J.; Wang, P.; Ke, P.; Wang, A. Microstructure and properties of (Cr:N)-DLC films deposited by a hybrid beam technique. *Surf. Coat. Technol.* **2015**, *261*, 398–403. [CrossRef]

33. Dai, W.; Wu, G.S.; Wang, A.Y. Structure and elastic recovery of Cr-C:H films deposited by a reactive magnetron sputtering technique. *Appl. Surf. Sci.* **2010**, *257*, 244–248. [CrossRef]

34. Hsu, J.S.; Tzeng, S.S.; Wu, Y.J. Influence of hydrogen on the mechanical properties and microstructure of DLC films synthesized by r.f.-PECVD. *Vacuum* **2009**, *83*, 622–624. [CrossRef]

35. Gou, W.; Chu, X.P.; Li, G.Q. Structure and properties of Cr-Containing hydrogenated amorphous carbon films synthesized by Filtered cathodic vacuum. *Plasma Process. Polym.* **2007**, *4*, S269–S272. [CrossRef]

36. Robertson, J. Diamond-like amorphous carbon. *Mater. Sci. Eng. R Rep.* **2002**, *37*, 129–281. [CrossRef]

37. Tay, B.; Cheng, Y.; Ding, X.; Lau, S.; Shi, X.; You, G.; Sheeja, D. Hard carbon nanocomposite films with low stress. *Diam. Relat. Mater.* **2001**, *10*, 1082–1087. [CrossRef]

38. Choi, J.; Ahn, H.; Lee, S.; Lee, K. Stress reduction behavior in metal-incorporated amorphous carbon films: First-principles approach. *J. Phys. Conf. Ser. IOP Publ.* **2006**, *29*, 155–158. [CrossRef]

39. Wang, A.Y.; Lee, K.-R.; Ahn, J.P.; Han, J.H. Structure and mechanical properties of W incorporated diamond-like carbon films prepared by a hybrid ion beam deposition technique. *Carbon* **2006**, *44*, 1826–1832. [CrossRef]

40. Dai, W.; Wu, G.; Wang, A. Preparation, characterization and properties of Cr-incorporated DLC films on magnesium alloy. *Diam. Relat. Mater.* **2010**, *19*, 1307–1315. [CrossRef]

41. Gassner, G.; Patscheider, J.; Mayrhofer, P.H.; Šturm, S.; Scheu, C.; Mitterer, C. Tribological properties of nanocomposite CrC x/aC: H thin films. *Tribol. Lett.* **2007**, *27*, 97–104. [CrossRef]

42. Michau, A.; Maury, F.; Schuster, F.; Boichot, R.; Pons, M. Evidence for a Cr metastable phase as a tracer in DLI-MOCVD chromium hard coatings usable in high temperature environment. *Appl. Surf. Sci.* **2017**, *422*, 198–206. [CrossRef]

43. Dai, W.; Liu, J.; Geng, D.; Guo, P.; Zheng, J.; Wang, Q. Microstructure and property of diamond-like carbon films with Al and Cr co-doping deposited using a hybrid beams system. *Appl. Surf. Sci.* **2016**, *388*, 503–509. [CrossRef]
44. Gayathri, S.; Kumar, N.; Krishnan, R.; Ravindran, T.; Dash, S.; Tyagi, A.; Sridharan, M. Influence of Cr content on the micro-structural and tribological properties of PLD grown nanocomposite DLC-Cr thin films. *Mater. Chem. Phys.* **2015**, *167*, 194–200. [CrossRef]
45. Xiao, Y.; Shi, W.; Luo, J.; Liao, Y. The tribological performance of TiN, WC/C and DLC coatings measured by the four-ball test. *Ceram. Int.* **2014**, *40*, 6919–6925. [CrossRef]

Article

The Influence of Pulling Up on Micropitting Location for Gears with Interference Fit Connections of Their Conical Surface

Layue Zhao [1,2], Yimin Shao [1,*], Minggang Du [2], Yang Yang [2] and Jixuan Bian [2]

1 State Key Laboratory of Mechanical Transmission, Chongqing University, Chongqing 400044, China; 15011080272@163.com
2 Science and Technology on Vehicle Transmission Laboratory, China North Vehicle Research Institute, Beijing 100072, China; mgdu@noveri.com.cn (M.D.); yangyang913@163.com (Y.Y.); bian110@sina.com (J.B.)
* Correspondence: shaoym_888@163.com

Received: 19 November 2020; Accepted: 8 December 2020; Published: 14 December 2020

Abstract: Micropitting is a surface fatigue phenomenon that occurs in Hertzian type of rolling and sliding contact that operates in elastohydrodynamic or boundary lubrication regimes and can progress both in terms of depth and extent. If micropitting continues to propagate, it may result in reducing gear tooth accuracy, increasing dynamic loads and noise. Eventually, it can develop into macropitting and other modes of gear failure such as flank initiated bending fatigue. Micropitting has become a particular problem in the gear surface fatigue. Usually micropitting initiates in the dedendum of the driver and driven at the asperities on the surface. However, the authors found for some gears with interference fit connections of their conical surface, micropitting on the pinion occurs in the addendum. This study attempted to find the reason using a 3D–TCA method based on ISO/TR 15144-1 to predict the micropitting and try to understand the key influence likely to affect micropitting location.

Keywords: micropitting; gear surface fatigue; dedendum; addendum

---

1. Introduction

Contact fatigue damage is a common mode of gear failure that is generally manifested as the initiation and progression of micropitting on the flank surface of gear teeth [1–10]. Micropitting is a phenomenon that occurs in Hertzian type of rolling and sliding contact that operates in elastohydrodynamic or boundary lubrication regimes [11,12]. Micropitting can affect all types of gears, it has been extensively studied by researchers [13–16] and is well known to gear designers [17]. Besides operating conditions such as load, speed, sliding, temperature, surface topography and specific film thickness, the chemical composition of lubrication strongly influences micropitting [18–25]. While these parameters are known to affect the performance of micropitting for a gear set, the subject area remains a topic of research. There has been a lot of research done on the impact of surface roughness, load, temperature, hardness and slide to roll ratio on micropitting [26–30]. The asperity contact stress between contact surfaces is one of the main reasons to promote micropits [25,26]. It is therefore generally considered that micropits can be prevented or lessened by reducing the roughness of the surface. Olver [30] found that the micropitting level was lessened by increasing the slide to roll ratio (SRR) but the wear rate did not change significantly.

Micropitting is characterized by the presence of fine surface pits and the occurrence of local plastic deformation and shallow surface cracks. Shallow cracks grow against the sliding direction on the gear flank, i.e., towards the pitch point on the driver tooth, and away from the pitch point on the driven tooth as illustrated in Figure 1. The cracks progress to form micropits which appears as a dull, matte surface to the observer, as shown in Figure 2.

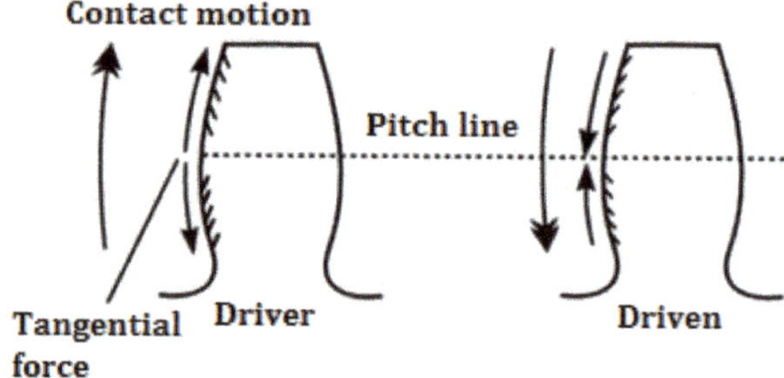

Figure 1. Illustration of direction of growth of micro cracks on the gear teeth from reference [19].

Figure 2. Gear tooth showing micropitting damage in the root region. (Design Unit, Newcastle University from reference [20]).

Micropitting can initiate during running and initial operation and then stop and stabilize. However, if micropitting continues to propagate, it may result in reduced gear tooth accuracy and increasing dynamic loads and noise. Eventually, it can develop into macropitting and other modes of gear failure such as flank-initiated bending fatigue. As a result of this, solutions avoiding micropitting of gear teeth are gaining interest with recent studies including the investigation of improved surface finishes, optimized microgeometry to minimize applied stress, modification of near surface residual stresses and the use of coatings. Zhang [31] investigated influence of shot-peening and surface finish on the fatigue of gears. A superfinished surface was found to be the most resistant to contact fatigue damage whereas shot peening resulted in a rougher tooth surface and a decreased contact fatigue life. Britton [32] used a special four-gear rig to determine gear tooth frictional losses and found that superfinishing resulted in a reduction of friction of typically 30 percent with correspondingly lower tooth surface temperatures under the same conditions of load and speed. Frazer [33] optimized micro-geometry to minimize applied stress for low friction losses. Moorthy [34] found that both BALINITs C and Nb–Scoated gears showed enhanced resistance to micro-pitting damage by removing localized stress concentration at microvalleys present on as ground gears.

Micropitting has become a particular problem in the gear surface fatigue area over the last 20 years. Remarkable developments in micropitting studies have been achieved recently by many researchers and engineers on both theoretical and experimental fields. Large amounts of investigations are yet to be further launched to thoroughly understand the micropitting mechanism. Morales Espejel [35] studied the occurrence of micropitting damage in gear teeth contacts. In his study, an existing general micropitting model which accounts for mixed lubrication conditions, stress history and fatigue damage accumulation, was adapted to deal with the transient contact conditions that exist during gear teeth meshing. The model considered the concurrent effects of surface fatigue and mild wear on the evolution of tooth surface roughness and therefore captured the complexities of damage accumulation on tooth flanks in a more realistic manner than hitherto possible. Clarke [36] made progress towards an understanding of the basic mechanism of micropitting in gears based on analysis of the contact mechanics and elastohydrodynamic lubrication (EHL) of gear tooth surfaces under realistic operating conditions. Results are presented which demonstrate the crucial influence of EHL film thickness in relation to roughness on predicted contact and near-surface fatigue. Hohn [37] studied the effects of lubricant temperature, circumferential speed and tooth flank roughness, and showed major influences on the micropitting resistance of gear wheels. The lubrication film thickness and the surface roughness were found to be dominant parameters. The chemical characteristics of the lubricant, i.e., the base oil and its additives, are also important factors. However, currently there is no standard evaluation for determining micropitting risk although ISO TR 15144-1 [11] provides guidance on minimum film thickness, linking this to micropitting risk. The prediction of micropitting remains a challenging problem and needs to be investigated further.

It was demonstrated [38] that the location of micropitting is close to the surface near the root of the teeth (dedendum). Brandão [39] developed a numerical model of surface-initiated failures and then verified it by applying an actual micropitting test on the carburized gears [40]. The results clearly showed that micropitting could be observed between the pitch line region and the tooth root after the loading period. Winkelmann [41] showed that the most common type of micropitting was wear and it mostly occurred in dedendum. However, authors found for some test gears connected with shafts by tapered hole interference fit (the gear geometry is shown in Table 1) in the Design Unit, Newcastle University, micropitting on the pinion occurs at the addendum (the tip region) as shown in Figure 3.

From Figure 3a, we can see that the biggest profile deviation used to represent micropitting occurs above pitch line (addendum). It can be seen from Figure 3b that the micropitting damage (the dull and grey part) is above the pitch line (addendum), which is very different from the usual phenomenon (micropitting in the root region), shown in Figure 2.

This study attempted to find the reason of micropitting occurring at addendum using an experimentally validated 3D–TCA method—Gear Analysis for Transmission Error and Stress (GATES) [42] based on ISO/TR 15144-1 [11] to predict the micropitting and tried to understand the key influences likely to affect micropitting location.

Table 1. Gear parameters of test gears with interference fit connection of conical surface.

Parameters	Pinion	Wheel	Parameters	Pinion	Wheel
Z	16	24	Number teeth spanned/K	4	4
m_n (mm)	3.9	3.9	$W_{k\,nom}$ (mm)	42.385	42.268
α_n (°)	20	20	$W_{k\,min}$ (mm)	42.28	42.17
β (°)	30	30	$W_{k\,max}$ (mm)	42.18	42.07
Hand of helix	left	right	grade	5	5
x	0.29	0	Centre distance a (mm)	91.5	91.5
h/m_n	2.4	2.4	Surface roughness R_a (um)	0.8	0.8
b (mm)	25	25	Nominal backlash (mm)	0.249	
d_a (mm)	82.12	115.88	Backlash-max (mm)	0.45	
d_f (mm)	63.4	97.16	Backlash-min (mm)	0.65	

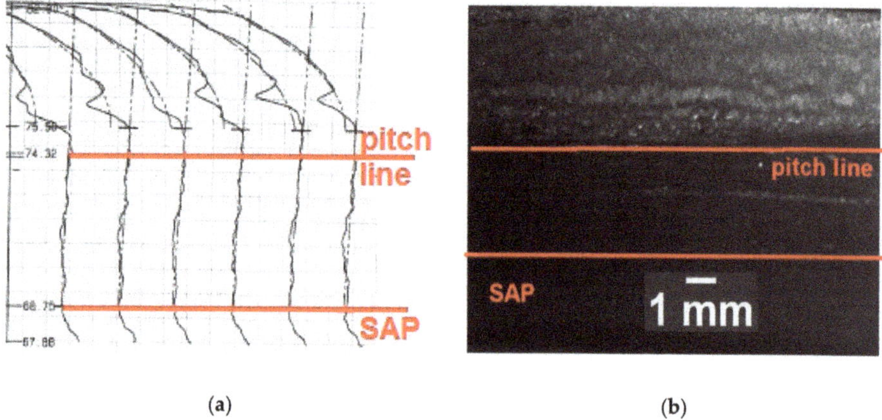

(a)　　　　　　　　　　　　　　　　　　　**(b)**

Figure 3. Micropitting on a pinion of test gears (**a**) Measurement data from gear measurement machine, (**b**) Optical image of the pinion flank replica.

2. Prediction of Micropitting

Micropitting is a very complicated surface fatigue and involves material, lubrication, speed, load, temperature, surface roughness, etc. Currently, there is no standard for determining micropitting. However, some methods attempt to predict micropitting. ISO/TR 15144-1:2014 [11] is one of the methods and provides principles for calculation of micropitting load capacity of cylindrical involute spur and helical gears with external teeth. The basis for the calculation of the micropitting load capacity of a gear set is the model of the minimum operating specific lubricant film thickness in the contact zone. Although the calculation of specific lubricant film thickness does not provide a direct method for assessing micropitting load capacity, it can serve as an evaluation criterion when applied as part of a suitable comparative procedure based on known gear performance.

In ISO/TR 15144-1:2014, the calculation of micropitting load capacity is based on the local specific film thickness $\lambda_{GF,Y}$ in the contact zone and the permissible specific film thickness λ_{GFP}. To account for micropitting load capacity, the safety factor S_λ according to Equation (1) is defined [11]:

$$S_\lambda = \frac{\lambda_{GF,min}}{\lambda_{GFP}} \geq S_{\lambda,min} \tag{1}$$

It is assumed that micropitting can occur when the safety factor is less than the minimum required safety factor. For the determination of the safety factor S_λ, the minimum specific film thickness has to be obtained from Equation (2):

$$\lambda_{GF,Y} = \frac{h_Y}{R_a} \tag{2}$$

where:

$$R_a = 0.5(R_{a1} + R_{a2}) \tag{3}$$

$$h_Y = 1600 \cdot \rho_{n,Y} \cdot G_M^{0.6} \cdot U_Y^{0.7} \cdot W_Y^{-0.13} \cdot S_{GF,Y}^{0.22} \tag{4}$$

where:

$$G_M = 10^6 \cdot \alpha_{\theta M} \cdot E_r \tag{5}$$

$$U_Y = \eta_{\theta M} \cdot \frac{v_{\Sigma,Y}}{2000 \cdot E_r \cdot \rho_{n,Y}} \tag{6}$$

$$W_Y = \frac{2 \cdot \pi \cdot P_{dyn,Y}^2}{E_r^2} \tag{7}$$

$$S_{GF,Y} = \frac{\alpha_{\theta B,Y} \cdot \eta_{\theta B,Y}}{\alpha_{\theta M} \cdot \eta_{\theta M}} \qquad (8)$$

The symbols are explained in the nomenclature section. There are no standard values for permissible specific film thickness. The determination of the permissible specific film thickness needs lots of experimental investigations and careful comparative studies, so it is very hard to obtain the exact permissible specific film thickness. Usually we just calculate the specific film thickness. Although the calculation of specific film thickness does not provide a direct method for assessing micropitting risk, it can serve as an evaluation criterion of comparative analysis.

In this study, we just calculate the specific film thickness to do some comparative analysis using GATES, an experimentally validated 3D–TCA method. This TCA method is based on ISO/TR 15144-1:2014 [11], using a full 3D FEA stiffness model to estimate the gear stiffness as the 1st stage and a 2nd stage tooth contact analysis to estimate the specific film thickness and other functional parameters. Its primary advantage is to predict specific film thickness by applying manufacturing deviations and specifying micro geometry corrections to consistent with the actual state of gears.

The procedures to calculate specific film thickness in GATES are as follows:

(1) Define the gear macro geometry, including tooth number, module, pressure angle, helix angle, face width, profile shift coefficient, tip diameter, root diameter and etc.
(2) Input rating parameters, including material data, duty cycle, quality and roughness, application data, misalignment, lubrication and etc.
(3) Input additional micropitting data, including lubricant viscosity at 40 °C and 100 °C, lubricant density at 15 °C, lubricant inlet temperature, lubricant method and etc. which are required in ISO/TR 15144-1:2014 [11].
(4) Check micropitting report in the theoretical analysis package Dontyne.
(5) Run GATES-TCA to check some results related with micropitting including sliding velocity, contact temperature and specific film thickness.

A specific film thickness result determined by GATES is shown in Figure 4. From Figure 4, we can see that the X-axis represents gear's facewidth, Y-axis represents the roll phase of driver, different colors represent different value of specific film thickness, the deepest red indicates the minimum specific film thickness. The area between SAP and pitch line of driver symbolizes the dedendum of driver and addendum of driven. The area between pitch line and EAP of driver symbolizes the addendum of driver and dedendum of driven. We can clearly see the amount and distribution of specific film thickness from GATES-3D result contour. We will use it to do the later comparative analysis.

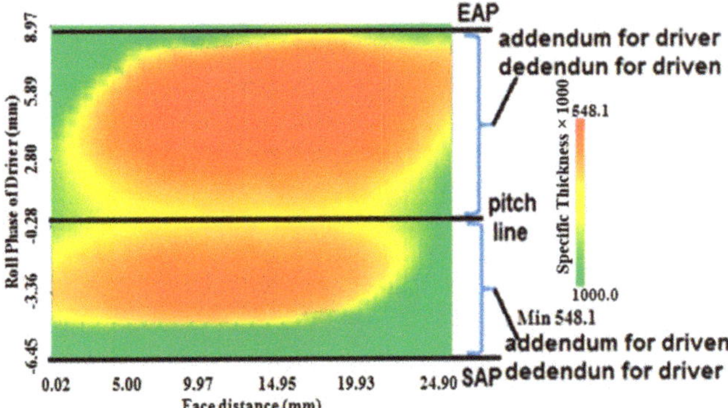

Figure 4. Example of specific film thickness determined by GATES.

3. Measurement Procedures

In order to find the reason of micropitting problem mentioned above, we did some tests of gears with all known parameters in the Design Unit, Newcastle University.

Before test, we measured the gears on a Klingelnberg P65 gear measuring machine (produced by Klingelnberg Co. Ltd., Ettlingen, Germany) that uses an involute generation measurement method to measure gear profile form deviations (Figure 5) to record the original profile/lead deviation and other parameters.

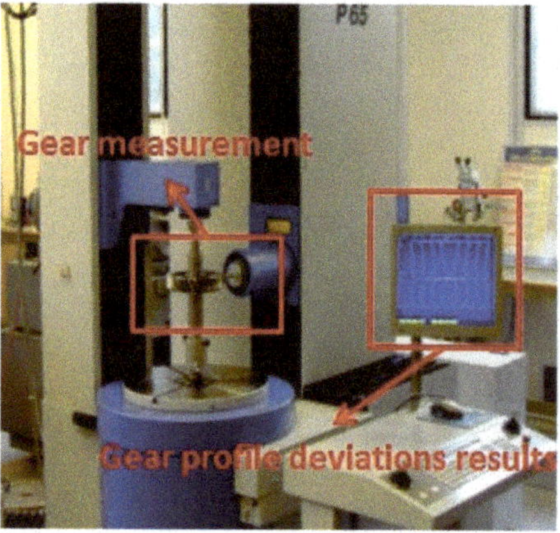

Figure 5. Gear measurement before testing on a gear measurement machine.

The test gears and shafts are connected by interference fit connection of their conical surface. Before test, gears must be pulled up on shaft to ensure accurate location and enough interference between gear bore and shaft must be guaranteed to transfer torque. The relationship between the distances of pulling up (x) and interference (δ) shown in Figure 6 is given by the following expression:

$$\delta = x \cdot \tan \theta \tag{9}$$

where θ is the angle of taper, $\tan\theta = 1/60$ for the test gears.

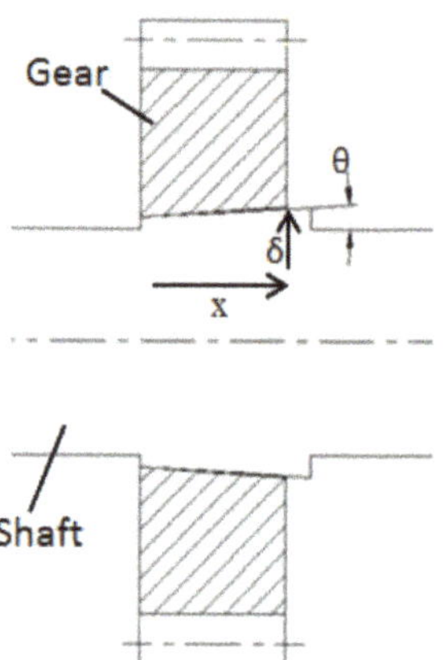

Figure 6. Relationship between pulling up x and interference δ.

In order to ensure that the connection of gear and shaft still has good tightness after multiple assembly and disassembly operations, the assembly and disassembly of gear with interference fit connection of conical surface adopts the hydraulic method, that is, injecting high-pressure oil between the mating surfaces, so as to increase the outer diameter of the hub housing and reduce the inner diameter of the shaft, it makes assembly and disassembly of the gear easily and also reduces the scratching on the mating surface. When using this method, it is necessary to open oil holes and grooves on the wheel hub, and special disassembly tools such as high-pressure oil pump, axial propeller, etc.

The procedures of pulling up gear on shaft showed in Figure 7 are as follows:

(1) Clean the assembly joint surface of gear and shaft with clean lubricating oil.
(2) Put the gear set on the shaft, turn it gently by hand, and push it up at the same time.
(3) Press the axial propulsion piston of the axial propeller into the guide cylinder, and then install the axial propeller on the shaft, and the axial propulsion piston can contact with the hub.
(4) Connect the oil pump and the pipeline to the shaft and the axial propeller, respectively.
(5) Start the high-pressure oil pump, inject oil into the oil hole on the shaft first, and then close the oil return valve after reaching a certain oil pressure to keep the oil pressure unchanged. The oil pressure value can be determined by GB/T 15755-1995 [43].
(6) Then, oil is injected into the axial thruster, and the axial propulsion piston pops out and the gear is pushed in. The distance of gear moving axially is the distance of pulling up.
(7) Open the return valve of the high-pressure oil pump; release the oil pressure of the high-pressure oil pump after releasing the oil pressure on the shaft.
(8) After installation, remove the oil injection pipeline and axial thruster.

(a)   (b)

Figure 7. Pulling up gear on shaft. (**a**) putting the gear set on the shaft; (**b**) injecting oil into the oil hole on the shaft and gear is pushed in

The steps of pulling out gears with interference fit connection of conical surface are as follows:

(1) In order to prevent the hub from popping out and damaging the parts, install the axial propeller on the shaft first, and leave a certain clearance between the hub and the axial press in device
(2) Connect the oil filling pipeline of the high-pressure oil pump to the oil filling hole of the shaft, and inject oil to a certain oil pressure (which can be determined by GB/T 15755-1995 [43]), and the rear wheel will pop out automatically
(3) Remove the oil injection pipeline and axial compressor.

To ensure the exact distance of pulling up are known, we measured the position of gears on shaft using an Endeavor 122010 coordinate measuring machine (CMM) (produced by Sheffield Company, Ruskin, FL, USA, and its measurement resolution is 0.1 μm) before and after pulling up. The measurement process is shown in Figure 8.

Figure 8. Measurement of gear location.4. Results and Discussion.

To make sure we measure the same location of gears, we made some marks on the gear hub and shaft as shown in Figure 8. After pulling up, we measured the gears on the Klingelnberg P65 gear measuring machine again to record the change of profile /lead deviation and other parameters.

4. Results and Discussion

4.1. Measurement Results

We chose one test gear pair to measurement. Gears were pulled up with very different distances, shown in Table 2.

Table 2. Distance of pulling up of test gears.

Pulling Up Stage	Pinion		Wheel	
	Z Position (mm)	Distance of Pulling Up (mm)	Z Position (mm)	Distance of Pulling Up (mm)
0	119.7489	-	120.1788	-
1	120.7193	0.9704	121.1268	0.948
2	121.511	1.7621	121.8763	1.6975
3	122.8564	3.1075	123.2744	3.0956
4	123.7481	3.9992	124.2064	4.0276

We measured the gears before and after pulling up in gear measuring machine Klingelnberg P65, the profile slope deviation ($f_{H\alpha}$) were showed in Table 3.

Table 3. Profile slope deviation of test gears before and after pulling up.

Pulling Up Stage	$f_{H\alpha}$ (um)	
	Pinion	Wheel
0	1.318	0.660
1	5.444	0.495
2	8.754	2.482
3	14.418	5.475
4	17.435	7.974

It is obvious that the pulling up distance influences the profile slope deviation ($f_{H\alpha}$) of the pinion and wheel, the profile slope deviation of the pinion is greater than that of the wheel with a similar pulling up distance and the bigger the pulling up, the greater the difference, which we can clearly see from Figure 9.

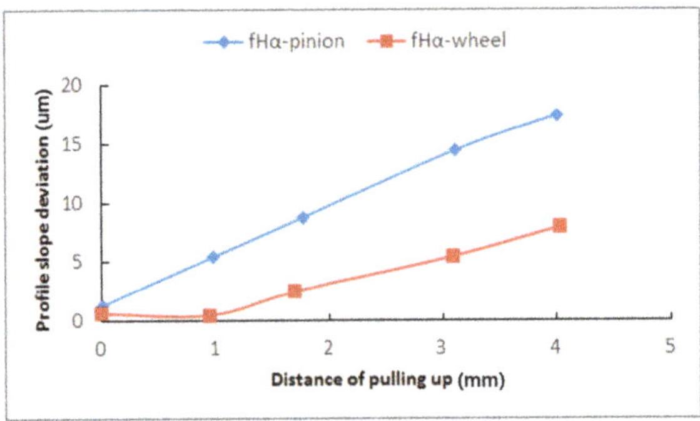

Figure 9. Relationship between pulling up and profile slope deviation for pinion and wheel.

In the process of pulling up, we know that it is difficult to control the pulling up distance so it is hard to ensure the distance of pulling up for pinion and wheel is the same. We found that when the distance of pulling up is equal to or slightly greater than 3.0 mm, the end of the gears will reach the edge of shaft shoulder, so for those test gears, the maximum distance of pulling up may be 3.0 mm or slightly greater than 3.0 mm. From the process of pulling up, we also found that it is easier to pull up the pinion than the wheel, so the distances of pulling up for the pinion and wheel are probably different, as represented in Table 4. We can see that the difference profile slope deviation between pinion and wheel are 9–10 um when the distance of pulling up is close to 3.0 mm, and the profile slope deviation of pinion is greater than wheel.

Table 4. Difference of profile slope deviation between pinion and wheel.

Distance of Pulling Up (mm)		$f_{H\alpha 1}$-$f_{H\alpha 2}$ (µm)
Pinion (mm)	Wheel (mm)	
0.9704	0.948	4.949
1.7621	1.6975	6.272
3.1075	3.0956	8.943
3.9992	4.0276	9.461

4.2. Simulation Results

We simulated test gears using GATES analysis based on measurement data showed in Table 3 to predict specific film thickness (which can be used to evaluate micropitting) at one load stage. Some results are shown in Figures 10 and 11.

From Figures 10 and 11, we can clearly see that the minimum specific film thickness of the pinion does change from dedendum to addendum when the pulling up distance changes from 0 mm to 3.1 mm.

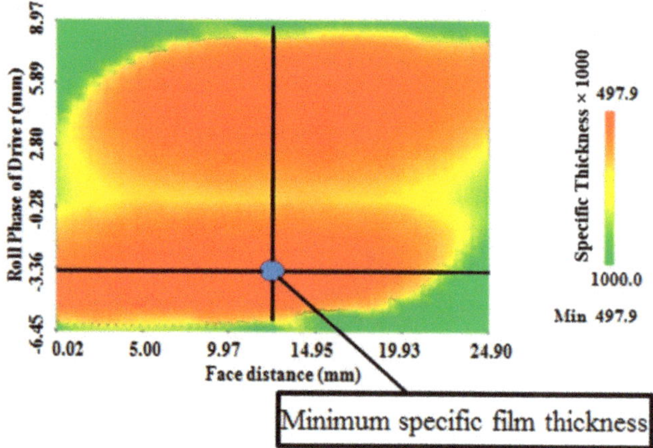

Figure 10. Specific film thickness of pinion for distance of pulling up equal to 0 mm.

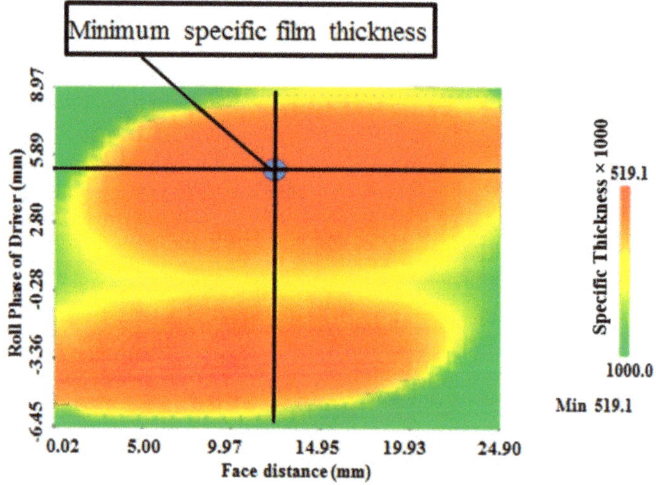

Figure 11. Specific film thickness of pinion for distance of pulling up equal to 3.1 mm.

We compared all the minimum specific film thicknesses on the dedendum and addendum at mid-facewidth which is comparable to measurement data and replica images. The results are listed in Table 5.

Table 5. Variation of specific film thickness with distance of pulling up changing.

Distance of Pulling Up (mm)	Minimum Specific Film Thickness $\lambda_{min} \times 1000$		$\dfrac{\lambda_{min-dedendum}}{\lambda_{min-addendum}}$
	Dedendum	Addendum	
0	497.9	527.6	0.944
1.0	522.3	523.3	0.998
1.7	526.5	521.8	1.009
3.1	531.6	519.1	1.024
4.0	532.6	519.1	1.026

From Table 5, we can see that the location of minimum specific film thickness will change from dedendum to addendum for the pinion when the pulling up distance is greater than 1.7. For this test gear pair, the distance of pulling up is usually 3.0 mm or slightly greater than 3.0 mm, so the location of minimum specific film thickness (micropiting) of the pinion may occur in the addendum.

4.3. Discussions

In order to understand the key influences likely to affect the location of micropitting, some other parameters were also investigated, including profile shift coefficient, start of tip relief and lead slope deviation. The results are as follows.

4.3.1. The Influence of Profile Shift Coefficient on Location of Micropitting

From Table 1, we can see that the biggest difference between test pinion and wheel is the profile shift coefficient ($x_1 = 0.29$, $x_2 = 0$). If the difference profile shift coefficient can lead to different micropitting results, we checked it taking four kinds of combination of profile shift coefficient. The results from GATES are shown in Table 6.

Table 6. Minimum specific film thickness for different profile shift coefficient.

Profile Shift Coefficient	Minimum Specific Film Thickness $\lambda_{min} \times 1000$		$\dfrac{\lambda_{min-dedendum}}{\lambda_{min-addendum}}$
	Dedendum	**Addendum**	
$x_1 = 0.29$, $x_2 = 0$	145	224	0.647
$x_1 = 0.3834$, $x_2 = 0$	223	243	0.918
$x_1 = 0.5$, $x_2 = 0$	250	248	1.008
$x_1 = 0.8$, $x_2 = 0$	320	258	1.240

Note: The lubricant which is used to calculate the micropitting is different from the above calculation. It is no problem if we just check the trend.

Misalignment and manufacturing deviation are not considered in this calculation. From Table 6, we can see that the different profile shift coefficients between pinion and wheel do influence the location of micropitting if the profile shift coefficient of the pinion is bigger enough than that of the wheel. For example, the minimum specific film thickness on the addendum is less than on the dedendum for the condition of $x_1 = 0.5$, $x_2 = 0$.

However, the profile shift coefficient of those test gears ($x_1 = 0.29$, $x_2 = 0$) is not enough to change the location of minimum specific film thickness. F. If the profile shift coefficient is the reason, and all gears would have the same results, so we can conclude that the current profile shift coefficient is not the reason for the uncommon micropitting location results.

4.3.2. The Influence of Start of Tip Relief on Location of Micropitting

For those test gears, tip reliefs are applied to reduce transmission error. The start of tip relief of pinion is at 75.58 mm diameter which is between pitch diameter (73.2 mm) and highest point of single tooth contact (HPSTC) diameter (79.572 mm), the start of the tip relief of the wheel is at 110.14 mm diameter which is also between pitch diameter (109.8 mm) and HPSTC diameter (113.665 mm). If the start of tip relief can affect the location of minimum specific film thickness, we can recalculated the micropitting by changing the start of tip relief to check the results.

From Table 7, we can see that no matter whether the start of tip relief is at the pitch diameter or HPSTC diameter or between them, the minimum specific film thickness is always at the dedendum, so the current start of tip relief also isn't the reason of the uncommon micropitting location results.

Table 7. Minimum specific film thickness for different start of tip relief.

Start of Tip Relief	Minimum Specific Film Thickness $\lambda_{min} \times 1000$		$\frac{\lambda_{min-dedendum}}{\lambda_{min-addendum}}$
	Dedendum	Addendum	
At pitch diameter	232.7	255.2	0.912
At HPSTC diameter	222.3	247.0	0.900
Between pitch diamete and HPSTC diameter	240.2	267.2	0.899

4.3.3. The Influence of Lead Slope Deviation on Location of Micropitting

From the above, we can see the differences of profile slope deviation between pinion and wheel influence the location of the minimum specific film thickness. To see if the differences of lead slope deviation ($f_{H\beta}$) have the same influence we defined different lead slope deviations to check the results, including $f_{H\beta1} = 5$ μm and $f_{H\beta2} = 0$, $f_{H\beta1} = 10$ μm and $f_{H\beta2} = 0$, $f_{H\beta1} = 15$ μm and $f_{H\beta2} = 0$, $f_{H\beta1} = 20$ μm and $f_{H\beta2} = 0$, $f_{H\beta1} = 0$ and $f_{H\beta2} = 5$ μm, $f_{H\beta1} = 0$ and $f_{H\beta2} = 10$ μm, $f_{H\beta1} = 0$ and $f_{H\beta2} = 15$ μm, $f_{H\beta1} = 0$ and $f_{H\beta2} = 20$ μm, where, $f_{H\beta1}$ is the lead slope deviation of pinion, $f_{H\beta2}$ is the lead slope deviation of wheel. The results of micropitting from GATES are shown in Table 8.

Table 8. Variation of minimum specific film thickness with different lead slope deviations.

Lead Slope Deviation (um)		Minimum Specific Film Thickness $\times 1000$		$\frac{\lambda_{min-dedendum}}{\lambda_{min-addendum}}$
Pinion	Wheel	Dedendum	Addendum	
0	0	360.6	365.6	0.986
5	0	361.3	364.8	0.990
10	0	362.0	364.8	0.992
15	0	362.8	364.7	0.995
20	0	363.6	364.7	0.997
0	5	359.9	365.4	0.985
0	10	359.3	365.2	0.984
0	15	358.4	365.1	0.982
0	20	357.6	365.2	0.979

From Table 8, it is obvious that the difference of lead slope deviation between the pinion and wheel cannot affect the location of the minimum specific film thickness and the minimum specific film thickness always occurs at the dedendum, so we can conclude that the lead slope deviation is not the reason for the uncommon micropitting location results.

For some test gears with interference fit connections of their conical surface, micropitting on the pinion occurs at the addendum, which is very different from the usual phenomenon (micropitting in the dedendum). We tried to find the reasons from all aspects we can think of, including profile slope deviation, profile shift coefficient, start of tip relief and lead slope deviation. From the calculation results, we can know that the profile shift coefficient, start of tip relief and lead slope deviation are not the possible reasons of the problem. From the results of minimum specific film thickness influenced by profile slope deviation which is caused by pulling up, we did find that if the profile slope deviation of pinion is much greater than that of the wheel, the position of minimum specific film thickness of pinion will be changed from the dedendum to the addendum, so the difference of profile slope deviation between the pinion and wheel caused by pulling up is a probable reason affecting the location of micropitting.

5. Conclusions

From the measurement and simulation presented in this study, the following conclusions can be drawn:

(1) Profile shift coefficient, start of tip relief and lead slope deviation cannot affect the location of micropitting.

(2) Pulling up gears with interference fit connection of their conical surface can affect the profile slope deviation.

(3) For the test gears with interference fit connection of the conical surface, profile slope deviations on pinion resulting from pulling up are always greater than those of the wheel and the bigger the pulling up, the greater this difference is.

(4) For those studied test gears, pulling up can lead to a 9–10 μm difference of profile slope deviation between the pinion and wheel.

(5) The difference of profile slope deviation between pinion and wheel may affect the micropitting location. If the profile slope deviation of the pinion is much greater than that of the wheel, the minimum specific film thickness on the pinion can change from the dedendum to addendum. This is a probable reason for the mcropitting on the test pinion with interference fit connection of the conical surface occurring at the addendum.

Author Contributions: Conceptualization, L.Z. and Y.S.; methodology, L.Z.; software, L.Z.; validation, L.Z., Y.S., M.D., Y.Y., and J.B.; formal analysis, M.D.; investigation, Y.Y.; resources, J.B.; data curation, L.Z.; writing—original draft preparation, L.Z.; writing—review and editing, M.D.; visualization, M.D.; supervision, Y.S.; project administration, L.Z. All authors have read and agreed to the published version of the manuscript.

Funding: This research received no external funding.

Acknowledgments: The authors gratefully acknowledge the support from the Design Unit, Newcastle University UK, which provided technical support for gear measurement.

Conflicts of Interest: The authors declare no conflict of interest. The funders had no role in the design of the study; in the collection, analyses, or interpretation of data; in the writing of the manuscript, or in the decision to publish the results.

Nomenclature

$\lambda_{GF,min}$	The minimum specific lubricant film thickness in the contact area
$\lambda_{GF,Y}$	The local specific lubricant film thickness
λ_{GFP}	The permissible specific lubricant film thickness
$S_{\lambda,min}$	The minimum required safety factor
R_a	The effective arithmetic mean roughness value, um
R_{a1}	The arithmetic mean roughness value of pinion, um
R_{a2}	The arithmetic mean roughness value of wheel, um
h_Y	The local lubricant film thickness, um
$\rho_{n,Y}$	The normal radius of relative curvature at point Y, mm
G_M	The material parameter
U_Y	The local velocity parameter
W_Y	The local load parameter
$S_{GF,Y}$	The local sliding parameter
$\alpha_{\theta M}$	The pressure-viscosity coefficient at bulk temperature, m^2/N
E_r	The reduced modulus of elasticity, N/mm^2
$\eta_{\theta M}$	The dynamic viscosity of the lubricant at bulk temperature, N·s/m^2
$v_{\Sigma,Y}$	Sum of tangential velocities, m/s
$P_{dyn,Y}$	The local Hertzian contact stress, N/mm^2
$\alpha_{\theta B,Y}$	The pressure-viscosity coefficient at local contact temperature, m^2/N
$\eta_{\theta B,Y}$	The dynamic viscosity at local contact temperature, N·s/m^2
$f_{H\alpha}$	The profile slope deviation, μm
$f_{H\alpha 1}$	The profile slope deviation of pinion, μm
$f_{H\alpha 2}$	The profile slope deviation of wheel, μm
x_1	The profile shift coefficient of pinion

x_2 The profile shift coefficient of wheel

$f_{H\beta 1}$ The lead slope deviation of pinion, μm

$f_{H\beta 2}$ The lead slope deviation of wheel, μm

References

1. Brandão, J.A.; Martins, R.C.; Seabra, J.H.O.; Castro, M.J. Calculation of gear tooth flank surface wear during an FZG micropitting test. *Wear* **2014**, *311*, 31–39. [CrossRef]
2. Sheng, S. *Wind Turbine Micropitting Workshop: A Recap*; National Renewable Energy Laboratory: Golden, CO, USA, 2010.
3. Clarke, A.; Weeks, I.J.J.; Snidle, R.W.; Evans, H.P. Running-in and micropitting behavior of steel surfaces under mixed lubrication conditions. *Tribol. Int.* **2016**, *101*, 59–68. [CrossRef]
4. Al-Mayali, M.F.; Hutt, S.; Sharif, K.J.; Clarke, A.; Evans, H.P. Experimental and Numerical Study of Micropitting Initiation in Real Rough Surfaces in a Micro-elastohydrodynamic Lubrication Regime. *Tribol. Lett.* **2018**, *66*, 150. [CrossRef] [PubMed]
5. Weibring, M.; Gondecki, L.; Tenberge, P. Simulation of fatigue failure on tooth flanks in consideration of pitting initiation and growth. *Tribol. Int.* **2019**, *131*, 299–307. [CrossRef]
6. Zhou, Y.; Zhu, C.; Gould, B.; Demas, N.G.; Liu, H.; Greco, A.C. The effect of contact severity on micropitting: Simulation and experiments. *Tribol. Int.* **2019**, *138*, 463–472. [CrossRef]
7. Evans, H.P.; Snidle, R.W.; Sharif, K.J.; Shaw, B.A.; Zhang, J. Analysis of Micro-Elastohydrodynamic Lubrication and Prediction of Surface Fatigue Damage in Micropitting Tests on Helical Gears. *J. Tribol.* **2012**, *135*, 011501. [CrossRef]
8. Oila, A. Micro-Pitting and Related Phenomena in Case-Carburised Gears. Ph.D. Thesis, Newcastle University, Newcastle upon Tyne, UK, 2003.
9. Hu, Y.-Z.; Zhu, D. A Full Numerical Solution to the Mixed Lubrication in Point Contacts. *J. Tribol.* **2000**, *122*, 1–9. [CrossRef]
10. D'Errico, F. Micro-pitting damage mechanism on hardened and tempered, nitrided and carburizing steels. *Metall. Ital.* **2012**, *104*, 5–11.
11. ISO/TR 15144-1. *Calculation of Micropitting Load Capacity of Cylindrical Spur and Helical Gears—Introduction and Basic Principles: 2014*; International Organization for Standardization: Geneva, Switzerland, 2014.
12. *Effect of Lubrication on Gear Surface Distress*; AGMA: Alexandria, VA, USA, 2003.
13. Berthe, D.; Flamand, L.; Foucher, D.; Godet, M. Micro-pitting in Hertzian contacts. *Trans. ASME J. Lubr. Tech.* **1980**, *102*, 478–489. [CrossRef]
14. Spikes, H.; Olver, A.; MacPherson, P. Wear in rolling contacts. *Wear* **1986**, *112*, 121–144. [CrossRef]
15. Zhou, R.S.; Cheng, H.S.; Mura, T. Micropitting in Rolling and Sliding Contact Under Mixed Lubrication. *J. Tribol.* **1989**, *111*, 605–613. [CrossRef]
16. Webster, M.N.; Norbart, C.J.J. An Experimental Investigation of Micropitting Using a Roller Disk Machine. *Tribol. Trans.* **1995**, *38*, 883–893. [CrossRef]
17. Shotter, B.A. Micropitting: Its characteristics and implications, in performance and testing of gear oils and transmission fluids. *Inst. Pet.* **1981**, *53*, 320–323.
18. Zhao, J.; Sadeghi, F.; Michael, H. Analysis of EHL Circular Contact Start Up: Part I—Mixed Contact Model with Pressure and Film Thickness Results. *J. Tribol.* **2001**, *123*, 67–74. [CrossRef]
19. O'Connor, B.M. The influence of additive chemistry on micropitting. *Gear Technol.* **2005**, *22*, 34–41.
20. Snidle, R.W.; Evans, H.P.; Alanou, M.; Holmes, M. Understanding Scuffing and Micropitting of Gears. In *The Control and Reduction of Wear in Military Platforms, Proceedings of the RTO AVT Specialists' Meeting on "The Control and Reduction of Wear in Military Platforms"; RTO-MP-AVT-109, Williamsburg, VA, USA, 7–9 June 2003*; NATO STO: Williamsburg, VA, USA, 2004.
21. Oila, A.; Bull, S. Assessment of the factors influencing micropitting in rolling/sliding contacts. *Wear* **2005**, *258*, 1510–1524. [CrossRef]
22. Roy, S.; Ooi, G.T.C.; Sundararajan, S. Effect of retained austenite on micropitting behavior of carburized AISI 8620 steel under boundary lubrication. *Mater.* **2018**, *3*, 192–201. [CrossRef]
23. Sun, Y.; Bailey, R. Effect of sliding conditions on micropitting behaviour of AISI 304 stainless steel in chloride containing solution. *Corros. Sci.* **2018**, *139*, 197–205. [CrossRef]

24. Cen, H.; Morina, A.; Neville, A. Effect of slide to roll ratio on the micropitting behaviour in rolling-sliding contacts lubricated with ZDDP-containing lubricants. *Tribol. Int.* **2018**, *122*, 210–217. [CrossRef]

25. Guangteng, G.; Spikes, H. An Experimental Study of Film Thickness in the Mixed Lubrication Regime. *Tribol. Ser.* **1997**, *32*, 159–166. [CrossRef]

26. Soltanahmadi, S.; Morina, A.; Van Eijk, M.C.P.; Nedelcu, I.; Neville, A. Investigation of the effect of a diamine-based friction modifier on micropitting and the properties of tribofilms in rolling-sliding contacts. *J. Phys. D Appl. Phys.* **2016**, *49*, 505302. [CrossRef]

27. Rycerz, P.; Olver, A.; Kadiric, A. Propagation of surface initiated rolling contact fatigue cracks in bearing steel. *Int. J. Fatigue* **2017**, *97*, 29–38. [CrossRef]

28. Maya-Johnson, S.; Santa, J.F.; Toro, A. Dry and lubricated wear of rail steel under rolling contact fatigue—Wear mechanisms and crack growth. *Wear* **2017**, 240–250. [CrossRef]

29. Hutt, S.; Clarke, A.; Evans, H. Generation of Acoustic Emission from the running-in and subsequent micropitting of a mixed-elastohydrodynamic contact. *Tribol. Int.* **2018**, *119*, 270–280. [CrossRef]

30. Lainé, E.; Olver, A.; Beveridge, T. Effect of lubricants on micropitting and wear. *Tribol. Int.* **2008**, *41*, 1049–1055. [CrossRef]

31. Zhang, J. Influence of Shot-Peening and Surface Finish on the Fatigue of Gears. Ph.D. Thesis, Newcastle University, Newcastle upon Tyne, UK, 2005.

32. Britton, R.D.; Elcoate, C.D.; Alanou, M.P.; Evans, H.P.; Snidle, R.W. Effect of Surface Finish on Gear Tooth Friction. *J. Tribol.* **1999**, *122*, 354–360. [CrossRef]

33. Frazer, R.C.; Shaw, B.A.; Palmer, D.; Fish, M. Optimizing Gear Geometry for Minimum Transmission Error, Mesh Friction, Losses and Scuffing Risk through Computer Aided Engineering. (AGMA Technical Paper 09FTM07). In Proceedings of the AGMA Fall Technical Meeting, Alexandria, VA, USA, 2009. Available online: https://www.geartechnology.com/articles/0810/Optimizing_Gear_Geometry_for_Minimum_Transmission_Error,_Mesh_Friction_Losses_and_Scuffing_Risk_Through_Computer-_Aided_Engineering/ (accessed on 8 August 2020).

34. Moorthy, V.; Shaw, B.A. An observation on the initiation of micro-pitting damage in a s-ground and coated gears during contact fatigue. *Wear* **2013**, *297*, 878–884. [CrossRef]

35. Morales-Espejel, G.; Rycerz, P.; Kadiric, A. Prediction of micropitting damage in gear teeth contacts considering the concurrent effects of surface fatigue and mild wear. *Wear* **2018**, 99–115. [CrossRef]

36. Clarke, A.; Evans, H.; Snidle, R. Understanding micropitting in gears. *Proc. Inst. Mech. Eng. Part C* **2015**, *230*, 1276–1289. [CrossRef]

37. Hohn, B.R.; Oster, P.; Schrade, U. Studies on the micropitting resistance of case-carburized gears—Industrial application of the new calculation method. *VDI Ber.* **2005**, *1904*, 1287–1307.

38. Snidle, R.W.; Evans, H.P. Some aspects of gear tribology. *Proc. Inst. Mech. Eng. Part C J. Mech. Eng. Sci.* **2008**, *223*, 103–141. [CrossRef]

39. Brandão, J.; Seabra, J.; Castro, J. Surface initiated tooth flank damage: Part I: Numerical model. *Wear* **2010**, *268*, 1–12. [CrossRef]

40. Brandão, J.A.; Seabra, J.H.O.; Castro, M.J. Surface initiated tooth flank damage. Part II: Prediction of micropitting initiation and mass loss. *Wear* **2010**, *268*, 13–22. [CrossRef]

41. Winkelmann, L. *Surface Roughness and Micropitting, Proceedings of the Wind Turbine Tribology Seminar, Bromfield, Golden, CO, USA, 2010*; National Renewable Energy Laboratory: Golden, CO, USA, 2010.

42. *Gear Production Suite Brochure*; Dontyne Systems: Newcastle upon Tyne, UK, 2018.

43. GB/T15755. *The Calculation and Selection of Cone Interference Fits*; Technical Committee for Standardization of Tolerances and Fits: Beijing, China, 1995.

Publisher's Note: MDPI stays neutral with regard to jurisdictional claims in published maps and institutional affiliations.

Article

Experimental Study on the Thickness-Dependent Hardness of SiO$_2$ Thin Films Using Nanoindentation

Weiguang Zhang [1,†], Jijun Li [1,2,*,†], Yongming Xing [1], Xiaomeng Nie [1], Fengchao Lang [1], Shiting Yang [1], Xiaohu Hou [3] and Chunwang Zhao [1,2]

[1] College of Science, Inner Mongolia University of Technology, Hohhot 010051, China; zhangwg@imut.edu.cn (W.Z.); xym@imut.edu.cn (Y.X.); niexiaomeng@163.com (X.N.); langfengchao@aliyun.com (F.L.); yang8191384@163.com (S.Y.); cwzhao@shmtu.edu.cn (C.Z.)

[2] College of Arts and Sciences, Shanghai Maritime University, Shanghai 201306, China

[3] Test Center, Inner Mongolia University, Hohhot 010051, China; houxiaohuhu@163.com

[*] Correspondence: jijunli@imut.edu.cn

[†] These authors contributed equally to this work and should be considered co-first author.

Abstract: SiO$_2$ thin films are widely used in micro-electro-mechanical systems, integrated circuits and optical thin film devices. Tremendous efforts have been devoted to studying the preparation technology and optical properties of SiO$_2$ thin films, but little attention has been paid to their mechanical properties. Herein, the surface morphology of the 500-nm-thick, 1000-nm-thick and 2000-nm-thick SiO$_2$ thin films on the Si substrates was observed by atomic force microscopy. The hardnesses of the three SiO$_2$ thin films with different thicknesses were investigated by nanoindentation technique, and the dependence of the hardness of the SiO$_2$ thin film with its thickness was analyzed. The results showed that the average grain size of SiO$_2$ thin film increased with increasing film thickness. For the three SiO$_2$ thin films with different thicknesses, the same relative penetration depth range of ~0.4–0.5 existed, above which the intrinsic hardness without substrate influence can be determined. The average intrinsic hardness of the SiO$_2$ thin film decreased with the increasing film thickness and average grain size, which showed the similar trend with the Hall-Petch type relationship.

Keywords: SiO$_2$ thin film; nanoindentation; surface morphology; load-penetration curve; hardness

Citation: Zhang, W.; Li, J.; Xing, Y.; Nie, X.; Lang, F.; Yang, S.; Hou, X.; Zhao, C. Experimental Study on the Thickness-Dependent Hardness of SiO$_2$ Thin Films Using Nanoindentation. *Coatings* **2021**, *11*, 23. https://dx.doi.org/10.3390/coatings11010023

Received: 8 December 2020
Accepted: 24 December 2020
Published: 27 December 2020

Publisher's Note: MDPI stays neutral with regard to jurisdictional claims in published maps and institutional affiliations.

1. Introduction

Due to its good chemical stability, optical properties, dielectric properties, abrasion resistance and corrosion resistance, silicon dioxide (SiO$_2$) thin films have received intensive attention within the technology and scientific community [1–9], and are widely used in micro-electro-mechanical systems, integrated circuits and optical devices [10–15]. Tremendous efforts have been devoted to studying the preparation technology and optical properties of the SiO$_2$ thin films [16–19], while little attention has been paid to their mechanical properties. In fact, the mechanical properties of SiO$_2$ thin films can affect their optical and electrical properties, and can also affect the production yield, serving time and reliability of the SiO$_2$ thin related devices [16–22]. Therefore, in order to optimize the design, and improve the performance, service time and reliability of the SiO$_2$ thin film-related devices, the mechanical properties of SiO$_2$ thin films should be studied carefully.

The nanoindentation technique has become an important means for characterizing the mechanical properties of films at micro- and nano-scale due to its high load resolution and displacement resolution [23–26]. Using the nanoindentation technique, we can determine many mechanical parameters such as hardness, elastic modulus, fracture toughness and so on [27–30]. In general, the penetration depth should be less than 10% of the film's thickness in order to avoid the substrate effect in the nanoindentation test of the thin films [31].

The mechanical behavior of the SiO$_2$ thin films has been studied by some researchers. Qin et al. [32] studied the influence of residual stress on the Young's modulus determination for SiO$_2$ thin film by the surface acoustic waves (SAWs), indicating which indicated that the influence of residual stress on the determination for SiO_2 thin film by SAWs is small, and it can be ignored or revised. Rakshit et al. [33] measured in situ the stress evolution in SiO_2 thin films in situ during electrochemical lithiation/delithiation cycling by monitoring the substrate curvature using a multi-beam optical sensing method, finding that upon lithiation, SiO_2 undergoes extensive inelastic deformation, with a peak compressive stress of 3.1 GPa, and upon delithiation the stress becomes tensile, with a peak stress of 0.7 GPa. Ho et al. [34] quantified the mechanical adhesion of SiO_2 thin films onto polymeric substrates by analyzing the SiO_2 buckle morphologies generated under compressive stress. The impacts of the mechanical properties of SiO_2 layers, as well as a surface pretreatment on adhesion, are shown. The interfacial toughnesses of both configurations are assessed using the Hutchinson and Suo model, which involves buckle dimensions determined in situ using an optical profilometer, and elastic modulus of the SiO_2 thin films, characterized by nanoindentation. The surface pretreatment led to initiation of buckling at a higher strain. The same trend was observed for a layer with a lower stiffness and residual stress. Wang et al. [35] studied the substrate effects on the mechanical properties of SiO_2 thin films deposited respectively on K9 and two different of $Y_3Al_5O_{12}$ (YAG) crystals by nanoindentation and nanoscratch tests, showing that the Young's moduli of all films were similar, the damage mechanisms of SiO_2 thin films on K9 and YAG were different, and the adhesive forces of the film on the films on YAG (100) and YAG (111) were much less than homologous film on K9. Simurka et al. [36] investigated the mechanical properties of amorphous SiO_2 films deposited on soda-line silicate float glass by reactive radio frequency magnetron sputtering at room temperature in dependence of the process pressure. As the pressure changed from 0.27 to 1.33 Pa, the residual compressive stresses in the deposited films varied in the range from 440 to 1 MPa. The hardness and reduced elastic modulus values followed the same trend and declined with the increase of process pressure from 8.5 to 2.2 GPa and from 73.7 to 30.9 GPa, respectively. However, these research works rarely involve investigation of the thickness-dependent hardness of SiO_2 thin films using the nanoindentation technique.

Here, we present an experimental investigation of the thickness-dependent hardness of SiO_2 thin films through nanoindentation technique and atomic force microscope (AFM). The surface morphology, load-penetration depth curves and hardnesses of the SiO_2 films with different thicknesses were evaluated.

2. Theoretical Approach

The nanoindentation measurements were analyzed using the Oliver-Pharr method [37,38]. This technique continuously monitors the load (at load as low as a few μN) and the penetration depth (down to a few nanometers) of an indenter, usually of a Berkovich type, as it is pushed into and withdrawn from the surface of the sample. The hardness of the material is generally determined by dividing the peak load by the contact area at the projected residual contact area. The unloading curve is fitted with a power-law relationship:

$$P = \alpha(h - h_f)^m,\tag{1}$$

where P is the penetration load of indenter, h is the penetration depth, h_f is the residual penetration depth after completely unloading, and α and m are fitting parameters. The contact depth, h_c, can be estimated from the load-penetration depth data as:

$$h_c = h_{max} - \varepsilon(P_{max}/S),\tag{2}$$

where h_{max} is the maximum penetration depth at the peak load P_{max}, ε is an indenter dependant constant, for a Berkovich indenter, $\varepsilon = 0.75$, S is the stiffness of the test material, and can be obtained from the initial unloading slope by evaluating the maximum load and maximum penetration depth, which is $S = dP/dh$. Hardness is defined as the resistance to local plastic deformation. It can be expressed as the maximum indentation load P_{max}, divided by the projected contact area A:

$$H = P_{max}/A, \tag{3}$$

where the projected contact area A is a function of the contact depth h_c. For an ideal Berkovich indenter, the relationship between the projected contact area A and the contact depth h_c is as follows:

$$A = 24.6h_c{}^2, \tag{4}$$

3. Experimental Details

The SiO_2 films with different thicknesses used in this study were prepared by thermal oxidation of p-type (100) silicon wafers [39]. The silicon wafers were cut into 20×10 mm^2 samples and cleaned to remove impurities and native oxide on the surface by using a conventional Radio Corporation of America (RCA) process and 1% HF. Then the cleaned samples were loaded into a quartz furnace at 600 °C and the temperature was ramped at a rate of 10 °C/min to reach an oxidation temperature of 1100 °C. This oxidation temperature was retained for 1 h, 3 h, and 10 h with water (H_2O) vapor constantly flowing into the furnace. The vapor was formed ex situ by heating a bubbler containing deionized water at 95 °C. It was carried into the furnace by nitrogen (N_2) carrier gas (250 mL/min), which was flown through the water. After the oxidation, a postoxidation annealing process was performed for 1 h at 1100 °C in ambient N_2. Thus, three SiO_2 films with thicknesses of 500, 1000 and 2000 nm were obtained.

The surface morphology and roughness of the three SiO_2 thin films with different thicknesses were examined by using CSPM4000 atomic force microscopy (AFM) (Being Nano-Instruments, Ltd., Guangzhou, China). AFM measurements were performed at room temperature (25 °C) and a relative humidity of 50% in air. The AFM was used in tapping mode with the scan frequency of 5 Hz between the tip and the surface [40]. The scanning area was 7×7 µm^2, and the AFM image size was 512×512 pixel^2. The measurements were repeated three times for each SiO_2 thin film on different scanning areas to validate the reproducibility of the data [41].

The mechanical properties of SiO_2 films were investigated using a three-sided Berkovich diamond tip with the G200 Nano indenter (Aglient, SantaClara, CA, USA) at room temperature. The G200 Nano indenter had a maximum of 500 mN and a maximum indentation depth greater than 500 µm. With a load resolution of 50 nN and displacement resolution less than 0.01 nm, it could accurately test the mechanical behavior of the material at both the micro and nano scale. The nanoindentation measurements were conducted using continuous stiffness measurement procedures. The indentation loads acted on the film surface at a constant strain rate of 0.05 s^{-1} with different maximum penetration depths. When the penetration depth reached the maximum value, the corresponding peak load was held for 10 s and then unloaded at the same strain rate. Three test points at each maximum penetration depth were performed on the samples to confirm the reliability and repeatability of the nanoindentation tests. For the sake of eliminating interactions among the test points, the spacing between each pair of neighboring test points was set to 100 µm.

4. Results and Discussion

4.1. Morphology of the SiO_2 Thin Films

Figure 1 illustrates the atomic force microscopy (AFM) two-dimension (2-D) and three-dimension (3-D) images of the 500-nm-thick, 1000-nm-thick and 2000-nm-thick SiO_2 thin films. The surface roughness and average grain sizes of the three SiO_2 thin films are also shown in Table 1. The values of the surface roughness (R_a), the root-mean-square (RMS) roughness and the height of irregularities at ten points (R_z) of the three SiO_2 thin films with different thickness were in the order of nanometer, which indicated that all three SiO_2 thin films have good smoothness, ensuring the accuracy and consistency of the nanoindentation experimental results.

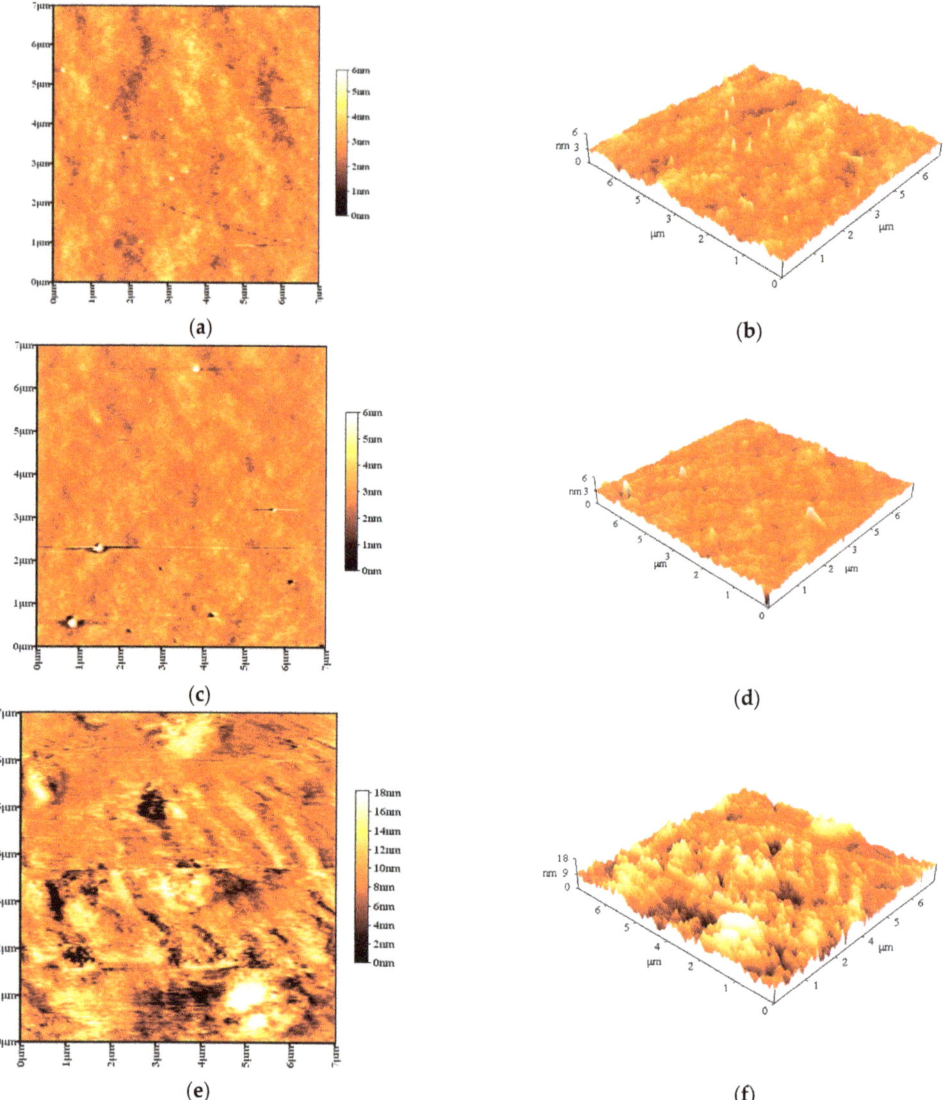

Figure 1. AFM images of the three SiO$_2$ thin films with different thicknesses: (**a**) 2-D AFM image of the 500-nm-thick SiO$_2$ thin film, (**b**) 3-D AFM image of the 500-nm-thick SiO$_2$ thin film, (**c**) 2-D AFM image of the 1000-nm-thick SiO$_2$ thin film, (**d**) 3-D AFM image of the 1000-nm-thick SiO$_2$ thin film, (**e**) 2-D AFM image of the 2000-nm-thick SiO$_2$ thin film, and (**f**) 3-D AFM image of the 2000-nm-thick SiO$_2$ thin film.

Table 1. Surface roughness and average grain sizes of the SiO$_2$ thin films with different thicknesses.

SiO$_2$ Film Thickness (nm)	R_a (nm)	RMS Roughness (nm)	R_z (nm)	Average Grain Size (nm)
500	2.84	3.23	3.35	60.7
1000	2.92	3.52	3.39	64.2
2000	3.00	3.36	10.96	66.9

The average grain sizes of the 500-, 1000- and 2000-nm-thick SiO_2 thin films were measured to be 60.7, 64.2 and 66.9 nm, respectively, and the corresponding change in the average grain size of the SiO_2 films with respect to film thickness is shown in Figure 2. It can be seen that the average grain size of SiO_2 thin film increases with increasing film thickness, which is because it is easier for atoms to migrate on the surface of the silicon wafer, and the islands are gradually connected according to the thin film growth diffusion principle.

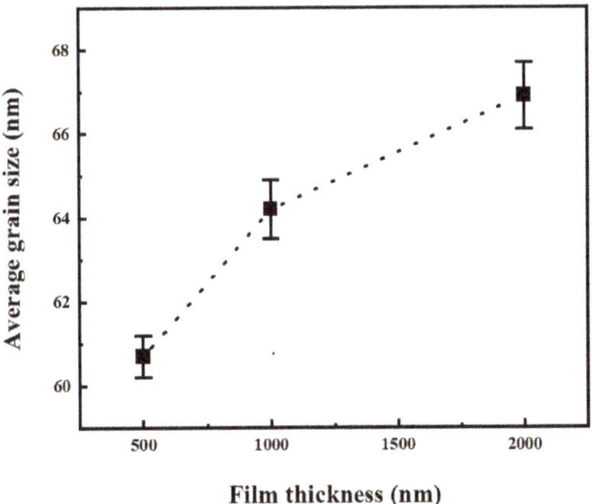

Figure 2. Change on average grain size of the SiO_2 film with respect to the film thickness.

4.2. Load-Penetration Depth Curves of the SiO_2 Thin Films

Figure 3a–c show the dependence of load (P) with the penetration depth (h) and relative penetration depth (h/t, where t is the thin film thickness) at different maximum penetration depths for the 500, 1000 and 2000-nm-thick SiO_2 thin films, respectively. It is observed that the P-h and P-h/t curves of the three test points at each maximum penetration depth h_{max} closely overlap for the three SiO_2 thin films. Therefore, the load-penetration depth and relative penetration depth curves demonstrate a small variation and low noise, the experimental data have good repeatability, reliability and accuracy, and the three SiO_2 thin films with different thicknesses have good uniformity.

As the h_{max} increases, the peak load P_{max} gradually increases. It can be seen that the P_{max} increased from 0.5 to 46.4 mN with the increasing h_{max} from 50 nm (1/10 of the film thickness) to 500 nm (film thickness) for the 500-nm-thick SiO_2 thin film, the P_{max} increased from 1.7 to 164.8 mN with the increasing h_{max} from 100 nm (1/10 of the film thickness) to 1000 nm (film thickness) for the 1000-nm-thick SiO_2 thin film, and the P_{max} increased from 5.7 to 642.1 mN with the increasing h_{max} from 200 nm (1/10 of the film thickness) to 2000 nm (film thickness) for the 2000-nm-thick SiO_2 thin film.

When the penetration depth reached the h_{max}, the corresponding peak load was held for 10 s and then unloaded. After unloading, penetration depth did not totally recover, and residual penetration depth remained, implying irreversible plastic deformation in SiO_2 thin films with different thicknesses during the loading process. Part of the loading–penetration depth curves of the three SiO_2 thin films with different thickness were extracted from Figure 3a–c, and shown in Figure 4. The loading curves of films with different thicknesses reflect the film's ability to obstruct its own deformation and resist external pressure. It can be seen that at the same penetration depth, the load was the highest on the SiO_2 film with a thickness of 500 nm, in the middle on the SiO_2 film with the thickness of 1000 nm, and the lowest on the SiO_2 film with the thickness of 2000 nm. Similarly, at the same load, the penetration depth was the lowest in the SiO_2 film with the thickness of 500 nm, in the middle in the SiO_2 film with the thickness of 1000 nm, and the highest in the SiO_2 film with the thickness of 2000 nm. Therefore, the SiO_2 film with the thickness of 500 nm has the highest resistance to external pressure, while the SiO_2 film with the thickness of 2000 nm has the lowest.

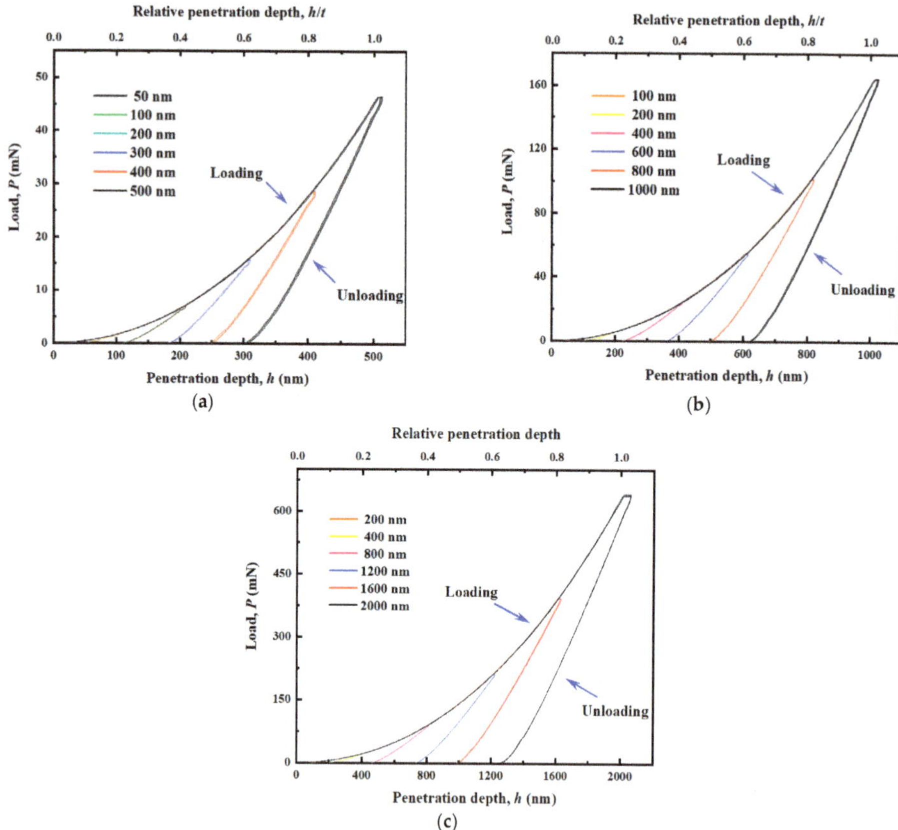

Figure 3. Load-penetration depth and relative penetration depth curves of the three SiO_2 thin films with different thicknesses: (**a**) 500-nm-thick SiO_2 thin film, (**b**) 1000-nm-thick SiO_2 thin film, and (**c**) 2000-nm-thick SiO_2 thin film.

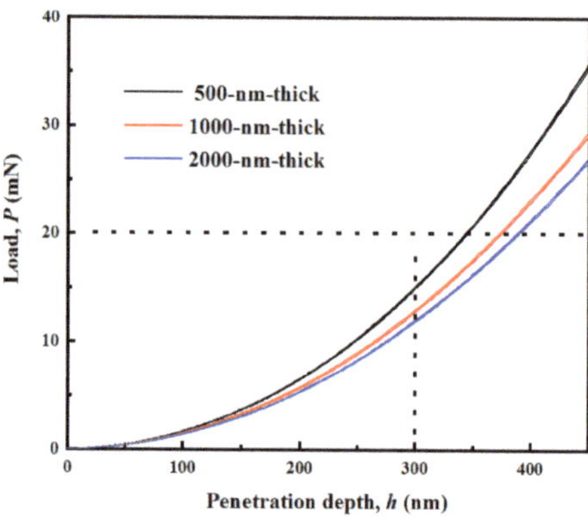

Figure 4. Part of loading-penetration depth curves of the three SiO_2 thin films with different thicknesses.

4.3. Hardness of the SiO$_2$ Thin Films

Figure 5a–c show the hardness–penetration depth curves of the 500-, 1000- and 2000-nm-thick SiO$_2$ thin films when the maximum penetration depths reached the corresponding film thicknesses, respectively. For the 500-, 1000- and 2000-nm-thick SiO$_2$ thin films, as the penetration depth was smaller than 100 nm, the data fluctuation was a bit high, which is probably associated with the resolution of the nanoindenter, the roughness of the sample surface, and ambient noise when searching for the initial contact position during the nanoindentation.

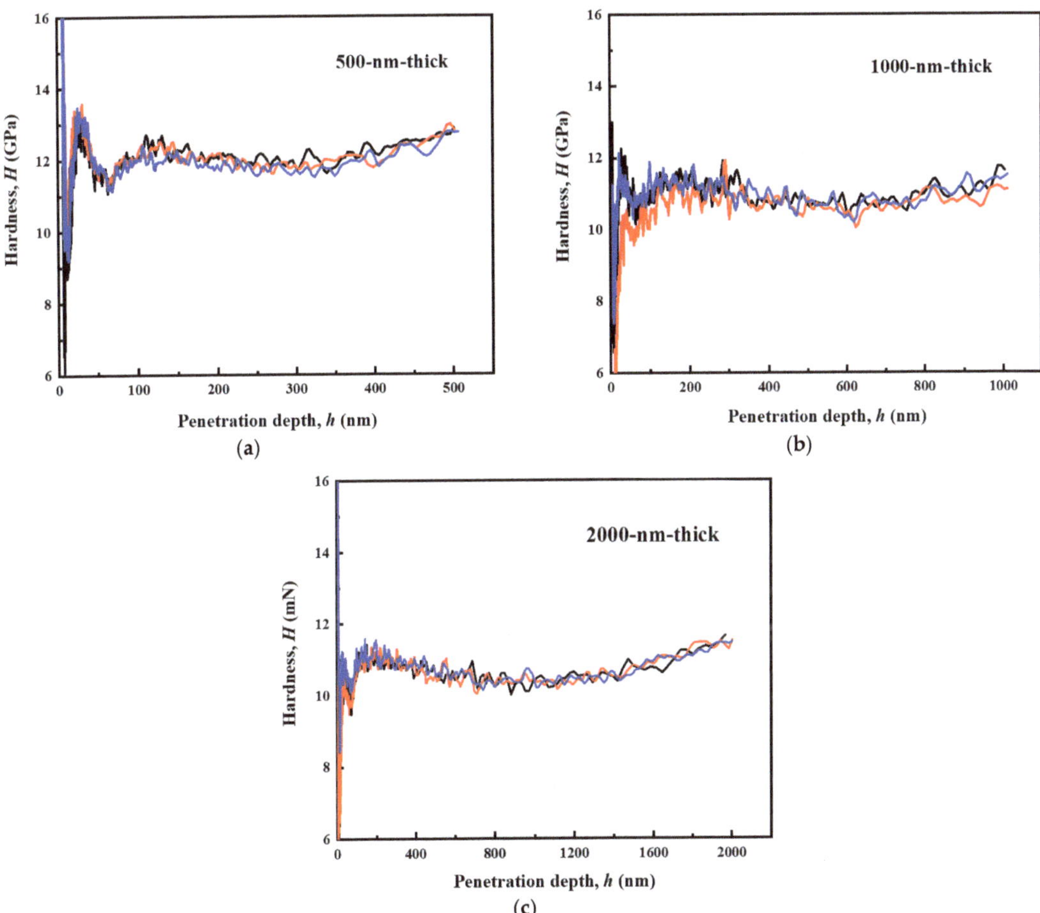

Figure 5. Hardness-penetration depth curves of the three SiO$_2$ thin films with different thicknesses: (**a**) 500-nm-thick SiO$_2$ thin film, (**b**) 1000-nm-thick SiO$_2$ thin film, and (**c**) 2000-nm-thick SiO$_2$ thin film.

In Figure 5a, as the penetration depth increased from 100 to 200 nm, the hardness of the 500-nm-thick SiO$_2$ thin film decreased slowly; as the penetration depth increased from 200 to 300 nm, the hardness tended to be stable; with increasing penetration depth more than 300 nm, the SiO$_2$ thin film was affected by the substrate, and the hardness increased gradually. Therefore, in order to obtain the intrinsic hardness of 500-nm-thick SiO$_2$ thin film without substrate influence, the penetration depth range should be ~200-300 nm. From Figure 5b,c, it can be seen that the hardnesses of the 1000- and 2000-nm-thick SiO$_2$ thin films exhibited the similar trends with that of 500-nm-thick SiO$_2$ thin film. To obtain the intrinsic hardnesses of 1000- and 2000-nm-thick SiO_2 thin films without substrate influence, the penetration depth regions should be ~400–500 nm and ~900–1000 nm, respectively.

To compare the hardnesses of 500-, 1000- and 2000-nm-thick SiO_2 thin films, the hardness was determined as the average of the values of three test points at each penetration depth. The corresponding hardness–relative penetration depth (H-h/t) curves of the 500-, 1000- and 2000-nm-thick SiO_2 thin films are shown in Figure 6. It can be seen that, for the three SiO_2 thin films with different thicknesses, the trends of hardness with the relative penetration depth were similar. The hardnesses of the three SiO_2 thin films with different film thicknesses were stable in the same relative penetration depth range of ~0.4–0.5. Therefore, for the three SiO_2 thin films, relative penetration depth ranges of ~0.4–0.5 existed, within which the intrinsic hardness could be determined without substrate influence.

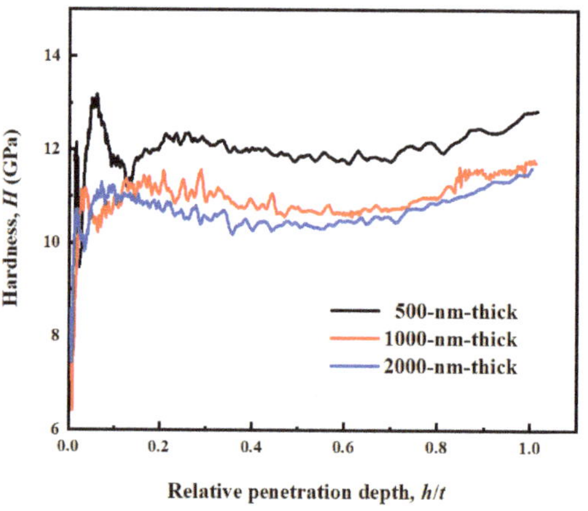

Figure 6. Hardness-relative penetration depth curves of SiO_2 thin films with different thin film thicknesses.

It should be noted that the general rule that relative penetration depth should not exceed 0.1 in order to obtain an intrinsic hardness of the thin film in nanoindentation is not a universal law. In the case of a soft thin film on a hard substrate, due to the confinement of the plastic deformed volume by lateral spreading within the soft thin film, the relative penetration depth has been found to be greater than 0.1 [42–44].

It also can be seen that the hardness of 500-nm-thick SiO_2 thin film was greater than that of 1000- and 2000-nm-thick SiO_2 thin films, and the hardness of 1000-nm-thick SiO_2 thin film was a little greater than that of 2000-nm-thick SiO_2 thin film. The average intrinsic hardnesses of the three SiO_2 thin films with different film thicknesses were evaluated in the relative penetration depth range of 0.4–0.5. In addition, the average intrinsic hardnesses of the 500-, 1000- and 2000-nm-thick SiO_2 thin films were 11.9, 10.7 and 10.4 GPa, respectively.

The dependence of the average intrinsic hardness of the SiO_2 thin film on film thickness is shown in Figure 7. It can be seen that the average intrinsic hardness of the SiO_2 thin film decreased with the increasing film thickness. As demonstrated in Figure 2, the average grain size of SiO_2 thin film increased with the increasing film thickness. Therefore, the average intrinsic hardness of the SiO_2 thin film decreased with the increasing average grain size (Figure 8). For the grain sizes of approximately 10–100 nm, Shockley partial dislocations or lattice dislocations that are nucleated in grain boundaries shear the grains and the dislocations are absorbed by the opposite grain boundaries. In this case, the hardness decreases with increasing grain size, namely, the Hall-Petch relationship holds [45–49]. Therefore, the results indicating that the average intrinsic hardness of the

SiO_2 thin film decreased with increasing film thickness exhibits a similar trend to that of the Hall-Petch type relationship.

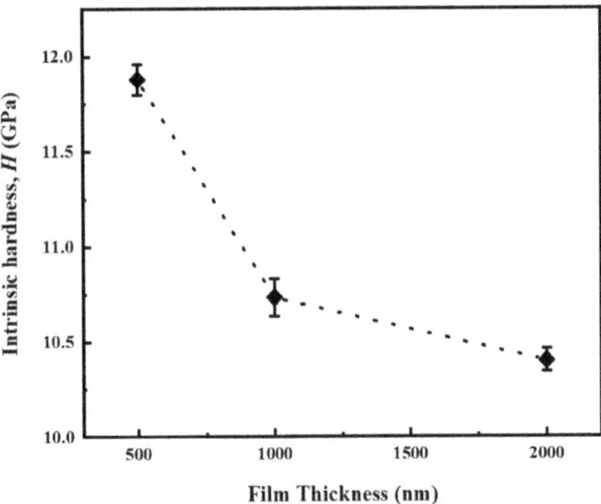

Figure 7. Dependence of the average intrinsic hardness of the SiO_2 thin film on the film thickness.

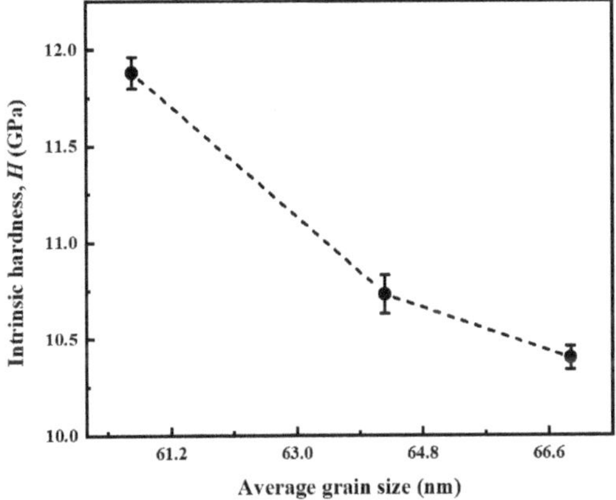

Figure 8. Change in average intrinsic hardness of the SiO_2 film with respect to the average grain size.

5. Conclusions

The surface morphology, load–penetration depth curves and the hardnesses of the 500-, 1000- and 2000-nm-thick SiO_2 thin films were investigated using the nanoindentation technique and force atomic force microscopy (AFM). The conclusions are as follows:

(1) The values of surface roughness parameters (R_a, RMS and R_z) of the three SiO_2 thin films with different thicknesses were in the order of nanometer, indicating that the three SiO_2 thin films with different thicknesses had good smoothness, ensuring the accuracy and consistency of the nanoindentation experimental results. The average grain sizes of the 500-, 1000- and 2000-nm-thick SiO_2 thin films were measured to be 60.7, 64.2 and 66.9 nm, respectively, indicating that the average grain size of SiO_2 thin film increased with increasing film thickness.

(2) The load–penetration depth curves of the three SiO_2 thin films with different thicknesses demonstrate small variation and low noise, the experimental data have good repeatability, reliability and accuracy, and the three SiO_2 thin films with different thicknesses have good uniformity. The SiO_2 film with the thickness of 500 nm had the highest resistance to external pressure, while the SiO_2 film with the thickness of

2000 nm had the lowest. Irreversible plastic deformation occurred in the three SiO_2 thin films with different thicknesses during the nanoindentation process.

(3) The average intrinsic hardnesses of the 500-, 1000- and 2000-nm-thick SiO_2 thin films were 11.9, 10.7 and 10.4 GPa, respectively. The average intrinsic hardness of the SiO_2 thin film decreased with increasing film thickness and average grain size, exhibiting a similar trend to the Hall-Petch type relationship.

Author Contributions: Conceptualization, J.L., W.Z., and Y.X.; methodology, J.L., W.Z., and X.N.; software, J.L., W.Z., and X.H.; validation, J.L., W.Z., and S.Y.; formal analysis, J.L. and W.Z.; investigation, J.L. and W.Z.; resources, J.L.; data curation, J.L., F.L., and W.Z.; writing—original draft preparation, J.L. and W.Z.; writing—review and editing, C.Z. and Y.X.; visualization, W.Z.; supervision, J.L. and Y.X.; project administration, J.L.; funding acquisition, J.L. and Y.X. All authors have read and agreed to the published version of the manuscript.

Funding: This research was funded by the National Science Foundation of China (Grant Nos. 11562016, 11972221, 12002174, 11762013, and 11762014), the Natural Science Foundation of Inner Mongolia Autonomous Region (Grant Nos. 2018MS01013 and 2013MS0107), and the Science Research Project of Inner Mongolia University of Technology (Grant Nos. X201713 and ZZ201812).

Data Availability Statement: Data is contained within the article.

Conflicts of Interest: The authors declare no conflict of interest.

References

1. He, P.; Wang, C.; Li, C.; Yang, J.; Qiu, F.; Wang, R.; Yang, Y. Optical properties of the low-energy Ge-implanted and annealed SiO_2 films. *Opt. Mater.* **2015**, *46*, 491–496. [CrossRef]
2. Bose, S.; Cunha, J.M.V.; Suresh, S.; Wild, J.D.; Lopes, T.S.; Barbosa, J.R.S.; Silva, R.; Borme, J.; Fernandes, P.A.; Vermang, B.; et al. Optical lithography patterning of SiO_2 layers for interface passivation of thin film solar cells. *Sol. RRL* **2018**, *2*, 1800212. [CrossRef]
3. Zukowski, P.; Koltunowicz, T.N.; Czarnacka, K.; Fedotov, A.; Tyschenko, I.E. Carrier transport and dielectric permittivity of SiO_2 films containing ion-beam synthesized InSb nanocrystals. *J. Alloys Compd.* **2020**, *846*, 156482. [CrossRef]
4. Huang, C.; Lin, Y. Surface properties of SiO_2 with and without H_2O_2 treatment as gate dielectrics for pentacene thin-film transistor applications. *Chem. Phys. Lett.* **2018**, *691*, 141–145. [CrossRef]
5. Li, W.; Mu, C. Charging effects of SiO_2 thin film on Si substrate irradiated by penetrating electron beam. *Micron* **2021**, *140*, 102961. [CrossRef] [PubMed]
6. Qin, F.; Jiang, L.; Long, P.; Xu, S.; Ling, Y.; Yu, Y.; Wei, G. Effect of deposition potential on preparation and corrosion resistance of SiO_2 film on copper. *Int. J. Electrochem. Sci.* **2020**, *15*, 6478–6487. [CrossRef]
7. Mazur, A.; Checmanowski, J.G.; Szczygiel, B. The effect of SiO_2 thin films deposited on the sol-gel method on steel 316L on its corrosion resistance and bioactivity in artificial blood solution. *Przem. Chem.* **2017**, *96*, 1183–1188.
8. Lu, J.; Xu, Y.; Zhang, Y.; Xu, X. The effects of SiO_2 coating on diamond abrasives in sol-gel tool for SiC substrate polishing. *Diam. Relat. Mater.* **2017**, *76*, 123–131. [CrossRef]
9. Chen, A.; Chen, Y.; Wang, Y.; Qin, J. Silica abrasives containing solid cores and mesoporous shells: Synthesis, characterization and polishing behavior for SiO_2 film. *J. Alloys Compd.* **2016**, *663*, 60–67. [CrossRef]
10. Vanheusden, K.; Warren, W.L.; Devine, R.A.B.; Fleetwood, D.M.; Schwank, J.R.; Shaneyfelt, M.R.; Winokur, P.S.; Lemnious, Z.J. Non-volatile memory device based on mobile protons in SiO_2 thin films. *Nature* **1997**, *386*, 587–589. [CrossRef]
11. Lin, Y.; Hung, C.; Huang, J.; Lin, S.; Chang, H. Electrical and surface properties of SiO_2 films modified by ultraviolet irradiation and used as gate dielectrics for pentacene thin-film transistor applications. *Chin. J. Phys.* **2019**, *61*, 248–254. [CrossRef]
12. Choi, Y.; Bae, S.; Kim, J.; Kim, E.; Hwang, H.; Park, J.; Yang, H.; Choi, E.; Hwang, J. Robust SiO_2 gate dielectric thin films prepared through plasma-enhanced atomic layer deposition involving di-sopropylamino silane (DIPAS) and oxygen plasma: Application to amorphous oxide thin film transistors. *Ceram. Int.* **2018**, *44*, 1556–1565. [CrossRef]
13. Wu, F.; Si, S.; Shi, T.; Zhao, X.; Liu, Q.; Liao, L.; Lv, H.; Long, S.; Liu, M. Negative differential resistance effect induced by metal ion implantation in SiO_2 film for multilevel RRAM application. *Nanotechnology* **2018**, *29*, 054001. [CrossRef] [PubMed]
14. Wostbrock, N.; Busani, T. Stress and refractive index control of SiO_2 thin films for suspended waveguides. *Nanomaterials* **2020**, *10*, 2105. [CrossRef] [PubMed]
15. Pfeiffer, K.; Shestaeva, S.; Bingel, A.; Munzert, P.; Ghazaryan, L.; Helvoirt, C.V.; Kessels, W.M.M.; Sanli, U.T.; Grévent, C.; Schütz, G.; et al. Comparative study of ALD SiO_2 thin films for optical applications. *Opt. Mater. Express* **2016**, *2*, 660–670. [CrossRef]
16. Ullah, A.; Wilke, H.; Memon, I.; Shen, Y.; Nguyen, D.T.; Woidt, C.; Hillmer, H. Stress relaxation in dual ion beam sputtered Nb_2O_5 and SiO_2 thin films: Application in a Fabry-Pérot filter array with 3D nanoimprinted cavities. *J. Micromech. Microeng.* **2015**, *25*, 055019. [CrossRef]
17. Sun, Q.; Sun, D.; Shang, H.; Yu, T.; Chang, L.; Sun, Z.; Xing, W.; Dong, M. Study of birefringence and stress distribution of SiO_2 film optical waveguide on silicon wafer. *Coatings* **2019**, *9*, 316. [CrossRef]

18. Hasunuma, R.; Kawamura, H.; Yamabe, K. Reliability factors of ultrathin dielectric films based on highly controlled SiO_2 films. *Jpn. J. Appl. Phys.* **2018**, *57*, 06KB05. [CrossRef]

19. Revilla, R.I.; Li, X.; Yang, Y.; Wang, C. Large electric field-enhanced-hardness effect in a SiO_2 film. *Sci. Rep.* **2014**, *4*, 4523. [CrossRef]

20. Jeffery, S.; Sofield, C.J.; Pethica, J.B. The influence of mechanical stress on the dielectric breakdown field strength of thin SiO_2 films. *Appl. Phys. Lett.* **1998**, *73*, 172–174. [CrossRef]

21. Yamabe, K.; Liao, K.; Minemura, H.; Murata, M. Nonuniform distribution of trapped charges in electron injection stressed SiO_2 films. *J. Electrochem. Soc.* **2001**, *148*, F9–F11. [CrossRef]

22. Yang, T.; Saraswat, K.C. Effect of physical stress on the degradation of thin SiO_2 films under electrical stress. *IEEE Trans. Electron Dev.* **2000**, *47*, 746–755. [CrossRef]

23. Guo, Y.; Staedler, T.; Fu, H.; Heuser, S.; Jiang, X. A novel way to quantitatively determine the mechanical properties of thin films from the initial-grown surface by nanoindentation. *Appl. Surf. Sci.* **2019**, *479*, 253–259. [CrossRef]

24. Broitman, E.; Tengdelius, L.; Hangen, U.D.; Lu, J.; Hultman, L.; Hoögberg, H. High-temperature nanoindentation of epitaxial ZrB_2 thin films. *Scripta Mater.* **2016**, *124*, 117–120. [CrossRef]

25. Wang, Y.; Zhang, J.; Wu, K.; Liu, G.; Kiener, D.; Sun, J. Nanoindentation creep behavior of Cu–Zr metallic glass films. *Mater. Res. Lett.* **2018**, *6*, 22–28. [CrossRef]

26. Haviar, S.; Kozák, T.; Meindlhumer, M.; Zitek, M.; Opatova, K.; Kučerová, L.; Keckes, J.; Zeman, P. Nanoindentation and microbending analyses of glassy and crystalline Zr(–Hf)–Cu thin-film alloys. *Surf. Coat. Technol.* **2020**, *399*, 126139. [CrossRef]

27. Xiao, K.; Li, J.; Wu, X.; Liu, H.; Huang, C.; Li, Y. Nanoindentation of thin graphdiyne films: Experiments and molecular dynamics simulation. *Carbon* **2019**, *144*, 72–80. [CrossRef]

28. Li, R.; Li, Y.; Yan, C.; Bao, Y.; Sun, H.; Hu, H.; Li, N. Thickness-dependent and tunable mechanical properties of $CaTiO_3$ dielectric thin films determined by nanoindentation technique. *Ceram. Int.* **2020**, *46*, 22643–22649. [CrossRef]

29. Grashchenko, A.S.; Kukushkin, S.A.; Osipov, A.V.; Redkov, A.V. Nanoindentation of GaN/SiC thin films on silicon substrate. *J. Phys. Chem. Solids* **2017**, *102*, 151–156. [CrossRef]

30. Ozmetin, A.E.; Sahin, O.; Ongun, E.; Kuru, M. Mechanical characterization of MgB_2 thin films using nanoindentation technique. *J. Alloys Compd.* **2015**, *619*, 262–266. [CrossRef]

31. Chen, J.; Bull, S.J. On the factors affecting the critical indenter penetration for measurement of coating hardness. *Vacuum* **2009**, *83*, 911–920. [CrossRef]

32. Qin, H.; Xiao, X.; Sui, X.; Qi, H. Influence of residual stress on the determination of Young's modulus for SiO_2 thin film by the surface acoustic waves. *Jpn. J. Appl. Phys.* **2019**, *58*, SHHG01. [CrossRef]

33. Rakshit, S.; Tripuraneni, R.; Nadimpalli, S.P.V. Real-time stress measurement in SiO_2 thin films during electrochemical lithiation/delithiation cycling. *Exp. Mech.* **2018**, *58*, 537–547. [CrossRef]

34. Ho, C.; Alexis, J.; Dalverny, O.; Balcaen, Y.; Faure, B. Mechanical adhesion of SiO_2 thin films onto polymeric substrates. *Surf. Eng.* **2018**, *35*, 1–6. [CrossRef]

35. Wang, H.; He, H.; Zhang, W. Substrate effects on the microstructure and mechanical properties of SiO_2 thin films. *J. Inorg. Mater.* **2013**, *28*, 653–658. [CrossRef]

36. Šimurka, L.; Čtvrtlik, R.; Tomaštík, J.; Bektaş, G.; Svoboda, J.; Bange, K. Mechanical and optical properties of SiO_2 thin films deposited on glass. *Chem. Pap.* **2018**, *72*, 2143–2151. [CrossRef]

37. Oliver, W.C.; Pharr, G.M. An improved technique for determining hardness and elastic modulus using load and displacement sensing indentation experiment. *J. Mater. Res.* **1992**, *7*, 1564–1583. [CrossRef]

38. Pharr, G.M.; Oliver, W.C. Measurement of thin film mechanical properties using nanoindentation. *Mrs Bulletin.* **1992**, *17*, 28–33. [CrossRef]

39. Poobalan, B.; Moon, J.H.; Kim, S.; Job, S.; Bahng, W.; Kang, I.H.; Kim, N.; Cheong, K. Investigation of SiO_2 film growth on 4H–SiC by direct thermal oxidation and postoxidation annealing techniques in HNO_3 & H_2O vapor at varied process durations. *Thin Solid Film.* **2014**, *570*, 138–149.

40. Sobola, D.; Ramazanov, S.; Konečný, M.; Orudzhev, F.; Kaspar, P.; Papež, N.; Knápek, A.; Potocek, M. Complementary SEM-AFM of swelling Bi–Fe–O film on HOPG substrate. *Materials* **2020**, *13*, 2402. [CrossRef]

41. Sobola, D.; Talu, S.; Sadovsky, P.; Papez, N.; Grmela, L. Application of AFM measurement and fractal analysis to study the surface the surface of natural optical structures. *Adv. Electr. Electron. Eng.* **2017**, *15*, 569–576. [CrossRef]

42. Gao, Q.; Jiang, H.; Li, M.; Lu, P.; Lai, X.; Li, X.; Liu, Y.; Song, C.; Han, G. Improved mechanical properties of SnO_2/F thin film by structural modification. *Ceram. Int.* **2014**, *40*, 2557–2564. [CrossRef]

43. Cai, X.; Bangert, H. Hardness measurements of thin films-determining the critical ratio of depth to thickness using FEM. *Thin Solid Film.* **1995**, *264*, 59–71. [CrossRef]

44. Bartali, R.; Vaccari, A.; Micheli, V.; Gottardi, G.; Pandiyan, R.; Collini, A.; Lori, P.; Coser, G.; Laidani, N. Critical relative indentation depth in carbon based thin films. *Prog. Nat. Sci.-Mater.* **2014**, *24*, 287–290. [CrossRef]

45. Hakamada, M.; Nakamoto, Y.; Matsumoto, H.; Iwasaki, H.; Chen, Y.; Kusuda, H.; Mabuchi, M. Relationship between hardness and grain size in electrodeposited copper films. *Mat. Sci. Eng. A* **2007**, *457*, 120–126. [CrossRef]

46. Yip, S. The strongest size. *Nature* **1998**, *391*, 532–533. [CrossRef]

47. Ellis, E.A.I.; Chmielus, M.; Han, S.; Baker, S.P. Effect of sputter pressure on microstructure and properties of β-Ta thin films. *Acta Mater.* **2020**, *183*, 504–513. [CrossRef]
48. Kumar, K.S.; Swygenhoven, H.V.; Suresh, S. Mechanical behavior of nanocrystalline metals and alloys. *Acta Mater.* **2003**, *51*, 5743–5774. [CrossRef]
49. Cheng, S.; Spencer, J.A.; Milligan, W.W. Strength and tension/compression asymmetry in nanostructured and ultrafine-grain metals. *Acta Mater.* **2003**, *51*, 4505–4518. [CrossRef]

coatings

Article

Nanoindentation Hardness and Practical Scratch Resistance in Mechanically Tunable Anti-Reflection Coatings

James J. Price [1], Tingge Xu [1], Binwei Zhang [1], Lin Lin [1], Karl W. Koch [1], Eric L. Null [1], Kevin B. Reiman [1], Charles A. Paulson [1], Chang-Gyu Kim [2], Sang-Yoon Oh [2], Jung-Keun Oh [2], Dong-Gun Moon [2], Jeong-Hong Oh [2], Alexandre Mayolet [1], Carlo Kosik Williams [1] and Shandon D. Hart [1,*]

[1] Corning Incorporated, Corning, NY 14831, USA; pricejj@corning.com (J.J.P.); xut2@corning.com (T.X.); zhangb4@corning.com (B.Z.); linl2@corning.com (L.L.); kochkw@corning.com (K.W.K.); nullel@corning.com (E.L.N.); reimankb@corning.com (K.B.R.); paulsonca@corning.com (C.A.P.); mayoleta@corning.com (A.M.); kosikwilca@corning.com (C.K.W.)

[2] Corning Incorporated, Asan-si, Chungcheongnam-do 336-725, Korea; changgyu.kim@corning.com (C.-G.K.); sangyoon.oh@corning.com (S.-Y.O.); jungkeun.oh@corning.com (J.-K.O.); donggun.moon@corning.com (D.-G.M.); jeonghong.oh@corning.com (J.-H.O.)

* Correspondence: hartsd@corning.com

Abstract: This work presents fundamental understanding of the correlation between nanoindentation hardness and practical scratch resistance for mechanically tunable anti-reflective (AR) hardcoatings. These coatings exhibit a unique design freedom, allowing quasi-continuous variation in the thickness of a central hardcoat layer in the multilayer design, with minimal impact on anti-reflective optical performance. This allows detailed study of anti-reflection coating durability based on variations in hardness vs. depth profiles, without the durability results being confounded by variations in optics. Finite element modeling is shown to be a useful tool for the design and analysis of hardness vs. depth profiles in these multilayer films. Using samples fabricated by reactive sputtering, nanoindentation hardness depth profiles were correlated with practical scratch resistance using three different scratch and abrasion test methods, simulating real world scratch events. Scratch depths from these experiments are shown to correlate to scratches observed in the field from consumer electronics devices with chemically strengthened glass covers. For high practical scratch resistance, coating designs with hardness >15 GPa maintained over depths of 200–800 nm were found to be particularly excellent, which is a substantially greater depth of high hardness than can be achieved using previously common AR coating designs.

Keywords: anti-reflection; nanoindentation; hardness; optical; interference; scratch; damage

Citation: Price, J.J.; Xu, T.; Zhang, B.; Lin, L.; Koch, K.W.; Null, E.L.; Reiman, K.B.; Paulson, C.A.; Kim, C.-G.; Oh, S.-Y.; et al. Nanoindentation Hardness and Practical Scratch Resistance in Mechanically Tunable Anti-Reflection Coatings. *Coatings* 2021, *11*, 213. https://doi.org/10.3390/coatings11020213

Academic Editor: Ben Beake

Received: 31 December 2020
Accepted: 8 February 2021
Published: 12 February 2021

Publisher's Note: MDPI stays neutral with regard to jurisdictional claims in published maps and institutional affiliations.

1. Introduction

Optical interference coatings utilize the physics of thin-film interference [1–4] to create unique optical properties that cannot be achieved using typical bulk materials. These coatings have been studied and applied for a broad variety of uses, including low reflection optics [4,5], high reflection mirrors [6–8], solar energy management [9], telecommunications [10], infrared (IR) sensors [11,12] and more. Despite this broad design flexibility and utility, optical interference coatings, especially anti-reflective (AR) interference coatings, have historically suffered from relatively low scratch or damage resistance, especially when compared to modern chemically strengthened glasses such as Corning® Gorilla® Glass. This relatively low durability remains true even for coatings using thin layers of materials which can have high hardness such as silicon nitride [13,14], because the design of these coatings relies on layers that are too thin (~0.1 micron or less) to protect the coating or the underlying substrate against common scratches in these types of applications. AR coatings are especially prone to high scratch visibility, due to high optical contrast (change in reflectivity) between the scratched and unscratched regions that readily occurs for scratches as shallow as ~50 nm.

In recent work, a new opto-mechanical design approach has been described enabling the production of anti-reflection coatings with very high hardness and scratch resistance [15,16]. Two key elements of this new coating design approach include: (1) A unique "modular" optical design approach which de-couples the optical performance of the coating from the thickness of a central high-hardness layer (the "hardcoat" layer or H-layer) in the multilayer design, and (2) selection of overall hardness vs. depth profiles of the coating to enable very high scratch resistance in aggressive real-world scratch events, such as those encountered in field trials of consumer electronics devices subject to daily handling. In this paper, the fundamental hardness and optical properties enabled by this tunable approach are further illuminated. Analysis of real-world field scratches from consumer electronics devices is used to establish a typical range of scratch depths in this field of use, which guides coating design choices for this application. Starting with a single-layer analysis, measured nanoindentation hardness vs. depth profiles are explained, considering a combination of substrate and coating effects. Extending this analysis to multilayer anti-reflection coatings, both modeling and experiments are used to show how this unique design approach enables the production of anti-reflection coatings with tunable hardness vs. depth profiles. Mechanical tuning through quasi-continuous thickness variation of the central hardcoat layer is shown to have minimal impact on the optical anti-reflection performance. This tunability allows for the ideal design to be selected from a broad range of options, depending on the level of hardness and scratch resistance that is desired for a particular end-use application. Finite element modeling is shown to be effective in predicting hardness vs. depth profiles in these multilayer anti-reflective hardcoatings. Finally, a high level of hardness maintained over extended depths is shown to correlate to scratch resistance in aggressive scratch testing designed to mimic severe real-world use conditions.

2. Materials and Methods

2.1. Coating Design and Fabrication

Anti-reflective, high-hardness coatings were designed using optical transfer matrix simulations [1], incorporating optical properties measured by spectroscopic ellipsometry for the experimentally fabricated coating materials. Coatings were made using reactive sputtering deposition processes [17–20]. These sputtering fabrication processes have been found to deliver desirable combinations of coating density, stress, hardness, optical absorption, layer thickness control, and deposition rates, enabling efficient manufacturing. Exemplary multilayer coatings are comprised of high hardness, high refractive index materials, such as SiN_x, SiO_xN_y, AlO_xN_y, or $SiAl_uO_xN_y$, layered with a low-index material, such as SiO_2. In this work, metallic aluminum or silicon sputtering targets were combined with non-reactive argon and reactive oxygen and nitrogen process gases in a sputtering chamber. For each coating material and sputtering process, a recipe development specific to each coating chamber geometry was required to achieve the desired targets of low optical extinction coefficient ($k < 0.001$ at 400 nm wavelength), high material hardness (>18 GPa for a single-material, 2-micron thick layer), and low or controlled film stress (0–400 MPa compressive). In order to achieve high hardness levels using SiN_x, SiO_xN_y, AlO_xN_y, or $SiAl_uO_xN_y$ materials, film compositions may include 0–20 atom% oxygen and 30–50 atom% nitrogen, with resulting refractive indices in the range of 1.9–2.1.

As an exemplary process, high hardness, high optical transparency SiO_xN_y film materials can be deposited by reactive magnetron sputtering from a silicon target [17,18]. Using a laboratory deposition chamber (ATC-2200 with A330-XP UHV magnetron sputtering source from AJA International (N. Scituate, MA, USA)) with 3" sputtering targets, example process conditions for SiO_xN_y deposition on glass substrates included 400 W RF power to the Si target supplied at 13.56 MHz, 30 sccm Ar gas flow, 40 sccm N_2 gas flow, 0.4 sccm O_2 gas flow, 1.5 mTorr process pressure, and a deposition temperature of 100–200 °C. These process conditions led to SiO_xN_y film materials with a refractive index of ~1.9 and measured hardness of >17 GPa, with a corresponding composition of ~48 atom% Si,

10 atom% O, and 42 atom% N. Using a similar process with O_2 gas flow set to zero, SiN_x or SiO_xN_y materials with only trace amounts of oxygen were deposited with a refractive index >2.0 and measured hardness >20 GPa. For films in this range of hardness, typical elastic modulus values were in the range of 200–240 GPa, resulting in H/E ratios of ~0.1.

These high-index, high-hardness materials were layered with a low-index material, typically SiO_2, to achieve the optical interference effects driving anti-reflection. It is necessary to have a low-index material component of the coating to achieve the desired interference effects, and materials with low refractive index typically have lower levels of hardness (due to the relationship of both hardness and refractive index to electron and bonding density in the materials). Lower-index, lower-hardness SiO_2 films can be made using similar deposition processes as those described above for SiN_x and SiO_xN_y, replacing N_2 gas flow with O_2 gas flow only. All coated samples, as well as the chemically strengthened glasses without hard coatings used for comparison, included a very thin (<10 nm) oleophobic fluorosilane layer on the top-most user-facing surface. These coatings are common on glasses in consumer electronics applications and tend to reduce the friction of the surface without having a significant effect on optics or hardness.

The opto-mechanical design strategy used to create high-hardness anti-reflective coatings [16] conceptually divides the coating into one or more optical interference layer regions, which perform the desired optical functions, and one or more hardcoat layer regions, which provide high levels of hardness at depths that are typical of real-world scratch events. In many cases the same material can be used for both the hardcoat region and the high-index component of the optical interference layer regions, while a softer low-index material is needed in the layered regions to create the desired optical interference effects. For anti-reflective coatings, multiple interference layer regions are employed for lower reflectance; the anti-reflective layers near the glass-to-hardcoat interface are also called "impedance matching" layers, to distinguish them from the AR layers adjacent to the user-facing, air-exposed surface. This combination of impedance-matching AR layers at the glass-to-hardcoat interface, together with the AR layers at the hardcoat-to-air interface, is what enables the uniquely tunable thickness (minimal optical change with thickness change) of the thick "hardcoat" layer.

2.2. Optical and Nanoindentation Measurements

Optical reflection and transmission spectroscopy were performed using an Agilent Cary 5000 UV-Vis-NIR spectrophotometer (Santa Clara, CA, USA) recording spectra with 2 nm wavelength step size and 4 nm resolution. The spectra were used to determine CIE L*a*b* color coordinates and perform reverse engineering of the designs. A Woollam M-2000 variable-angle spectroscopic ellipsometer, varying the input angle of the measurement from 45° to 80° and scanning the wavelength from 250 to 1680 nm, was used to analyze film thickness and refractive index. These measurements were used to determine single-layer film material dispersion and deposition rates for individual processes, as well as to reverse-engineer the refractive index and thickness for film stacks consisting of multiple layers.

Coating hardness was measured via nanoindentation using a Berkovich diamond tip on a KLA Instruments G200 Nano Indenter (Milpitas, CA, USA). Hardness was extracted using the widely accepted Oliver–Pharr approach [21,22], combined with a special technique known as the continuous stiffness measurement [23]. The continuous stiffness measurement technique superimposed a small (+/−1 nm) oscillatory signal at 52 Hz during the loading cycle, which enabled the hardness to be measured continuously as a function of indentation depth. The system can reach a maximum indentation depth of 500 µm (resolution of 0.2 nm) and a maximum load of 630 mN (resolution of 50 nN). A Berkovich indenter tip, made from a single crystal diamond, was used in this investigation. A constant indentation strain rate of $0.05\ \mathrm{s}^{-1}$ was maintained during the loading stage. The amplitude of harmonic oscillation was 1 nm. Nanoindentations at 10 different locations were conducted on each sample, with typical standard deviations of hardness in the range of 0.3–0.5 GPa.

Analyses were carried out to determine the mechanical properties (hardness and modulus) based on the contact mechanics analysis of nanoindentation load-depth relationships, which has been well established by Oliver and Pharr [21,22].

The indentation hardness (H) is obtained using

$$H = \frac{P_{max}}{A_c} \qquad (1)$$

where P_{max} is the maximum nanoindentation load, A_c is the contact area corresponding to the contact depth (h_c) at the maximum load.

The contact stiffness $S = dP/dh$ is calculated from the slope of the initial unloading cure, and the reduced modulus is obtained by,

$$E_r = \frac{1}{2\beta}\frac{\sqrt{\pi}}{\sqrt{A_c}}\frac{dP}{dh} \qquad (2)$$

where β is the indenter shape correction factor [21]. The reduced modulus E_r is related to the specimen's Young's modulus by,

$$1/E_r = (1 - v_s^2)/E_s + (1 - v_i^2)/E_i \qquad (3)$$

where E_s and v_s are the Young's modulus and Poisson's ratio of the specimen, respectively, while E_i and v_i are the Young's modulus and Poisson's ratio of the diamond Berkovich indenter tip, respectively.

The contact depth h_c is given by,

$$h_c = h_{max} - \varepsilon\frac{P_{max}}{S} \qquad (4)$$

Oliver and Pharr found that experimental results on standard specimens that indicated $\varepsilon = 0.75$ best accounts for the material behavior [21]. The area function was calibrated on a reference sample of fused silica. The area function, which accounts for the non-ideal shape of a real indenter, was fitted into the following polynomial,

$$A_c = 25.746h_c^2 + 1383.2h_c - 6942.11h_c^{1/2} + 6670.6h_c^{1/4} \qquad (5)$$

where the contact depth h_c has a unit of nm. Since the continuous stiffness measurement (CSM) method imposes a harmonic displacement (1 nm) during loading, the unloading stiffness can be obtained continuously, and thus, Young's modulus and hardness as a function of nanoindentation depth can be obtained based on the Oliver–Pharr analysis shown above.

2.3. Scratch Resistance Tests

In order to establish correlations between hardness vs. depth profiles and practical scratch resistance, three different controlled scratch tests were employed. These scratch tests were chosen because they create scratches with similar morphology and depth to the most frequent scratches observed in field studies of chemically strengthened glass covers for consumer electronics devices. The first "nano-scratch" test involves controlled scratching with a diamond tip using a KLA G200 nano indenter with scratch option. Using a Berkovich diamond tip, a constant load scratch of 500 μm length is applied at 50 μm/s velocity. The resulting scratch depth is then measured using the Berkovich tip as a surface profilometer by applying an ultra-light load of 50 micro-Newtons. This method of depth profiling has been found to give values close to an atomic force microscope (AFM), in the depth ranges of interest. The diamond scratch test gives precise and repeatable results, however, most real-world scratch events do not involve an abrasive as hard as diamond, so other methods were also evaluated. A second test method evaluated was reciprocating abrasion with 400 grit Al_2O_3 sandpaper from Kovax (Tokyo, Japan), using a 700 g load applied over a ~1 cm × 1 cm contact area for 50 reciprocating cycles (40 cycles/min, 1.5 inch long stroke length). This test method is more properly categorized as abrasion or wear, due to the reciprocating cycles, which are not typically observed in real-world scratch events. However, this method was chosen because of its relevance to surface durability and its ability to create a semi-uniform area of multiple scratches which can be quantified using light scattering methods. Light scattering is particularly relevant to optical coatings such as anti-reflection coatings, and was quantified using a reflective scattered light measurement with specular component excluded (SCE) measurement using a Konica-Minolta CM700D with a 6 mm diameter aperture [24,25]. Finally, a garnet scratch test method was evaluated consisting of a single pass with 150 grit garnet sandpaper, with 1 kg applied load over a ~0.6 cm diameter contact area. This garnet scratch test is the closest to simulating real-world scratch events, due to the single-pass nature of the test and the intermediate hardness of garnet, which is moderately harder than common real-world abradants such as sand. The garnet scratch test yields relatively sparse and random scratches which are difficult to analyze using light scattering, so this test was quantified using an atomic force microscope (AFM) measurement of each scratch that was visible to the naked eye, in the center of the scratch path.

3. Results

3.1. Field Analysis of Real-World Scratch Events

To establish a baseline target of real-world scratch events, the consumer electronics industry was chosen as our application focus. Consumer touch-screen devices with chemically strengthened glass covers were subjected to typical daily usage handling by actual consumers in the field, and were then characterized for the type, depth, and severity of observed scratches. About 75% of the observed scratches were characterized as single-event, groove-like micro-ductile type scratches [26], while about 25% were associated with some level of cracking or chipping. Focusing on the most frequently observed micro-ductile scratches, the scratch depths of these were measured using atomic force microscopy (AFM). The measured distribution of scratch depths is shown in Figure 1. The majority of the measured micro-ductile scratch depths are in the 100–500 nm depth range. This suggests that coating or surface hardness in this depth range is of particular importance in preventing these types of frequent real-world scratch events.

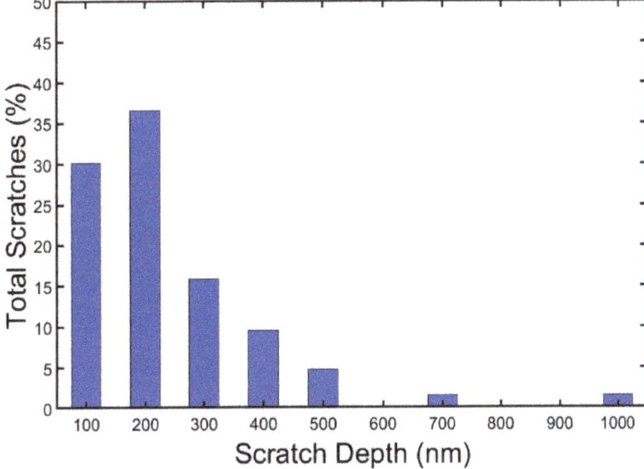

Figure 1. Measured scratch depths of micro-ductile scratches from field study of consumer electronics devices with chemically strengthened glass covers. The majority of observed scratch depths are in the 100–500 nm depth range.

3.2. Single-Layer Hardness Profiles vs. Scratch Depth

Informed by the scratch depths of interest from Figure 1, single-layer hardcoatings of varying thickness were studied for initial understanding of their effectiveness in suppressing scratches in this depth range. Considering the fundamentals of the nanoindentation hardness measurement, it is important to keep in mind how the indenter interacts with a coated material. As shown schematically in Figure 2, the region of permanent deformation (plastic zone) is associated with the hardness of the material. However, as shown in Figure 2, an elastic stress field extends well beyond this region of permanent deformation, meaning the indenter has an interaction field that can be considerably larger than the thickness of a thin film. As the indentation depth increases, the apparent hardness and modulus are influenced by stress field interactions with the substrate. The modulus is influenced by the substrate almost immediately, as the indenter begins to penetrate the sample surface since the region of elastic deformation extends far enough to interact with the substrate, even upon initial shallow indentation. The region of plastic deformation has a smaller extent than the region of elastic deformation. This has the result that substrate influence on hardness occurs at deeper indentation depths, since deeper penetration is needed to extend the region of plastic deformation that influences hardness to interact with the substrate. A transition where the hardness of the substrate begins to have an influence is typically observed when the indentation depth is ~30% of the coating thickness. Importantly, these volumetric interactions are related to the fundamental mechanics of contact damage, and can be expected to be similarly present in real-world scratching events. This means that the practical scratch resistance of a hardcoated surface may not only depend on the hardness of a thin film layer, but also on the hardness vs. depth profile and the interaction of multiple layers, up to—and potentially including—the substrate on which the coating is formed.

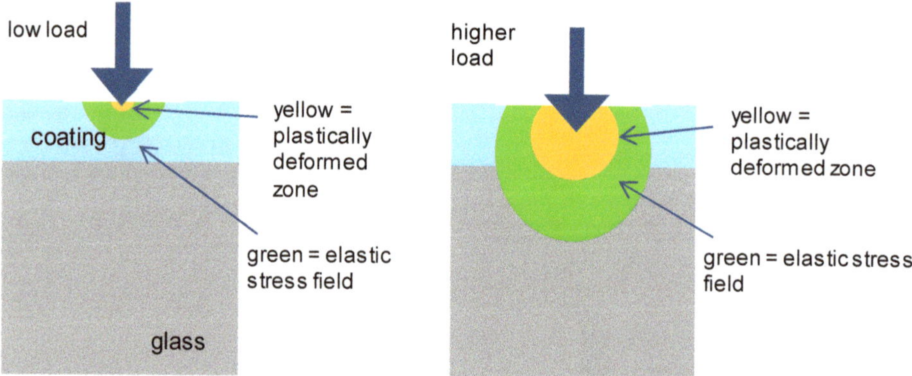

Figure 2. Schematic illustration of contact mechanics in nanoindentation testing.

Figure 3a shows how these interactions impact the measured hardness vs. depth profiles for a series of single-layer SiO_xN_y films of varying thickness coated on aluminosilicate glass substrates. At shallow depths <~100 nm, the hardness response requires a certain minimum load to develop full plasticity during the indentation process, causing the hardness to show a generally increasing trend at these shallow depths. The hardness then reaches a "plateau" and begins to drop gradually as the substrate interaction effects take on a larger role at deeper indentation depths. In Figure 3, the thin (700 nm), medium (1500 nm), and thick (2000 nm w ~20 GPa H) films were all deposited using the same process conditions, but the measured apparent hardness profiles of these three films vary substantially due to their differing thickness. The 700 nm film has a lower peak hardness, which is largely due to the aforementioned effects which can suppress measured hardness at both shallow depths and deeper depths. It is likely that these effects prevent us from measuring the "true" film hardness of this 700 nm coating. For the 1500 nm and 2000 nm films, a plateau occurs at approximately the same hardness level near ~20 GPa. This suggests that ~20 GPa is near the "true" hardness of these films, and that this true hardness can only be measured when the film is sufficiently thick. For comparison, the hardness of typical chemically strengthened glass without coatings is approximately ~8 GPa. The final thick sample (2000 nm w ~23 GPa H) was deposited using tuned process conditions to further maximize hardness, such as slightly lowering deposition pressure, slightly increasing deposition plasma energy, and small reductions in oxygen/increased nitrogen levels. While these changes can have a negative impact on other properties such as optical absorption and film stress, these conditions were tested here to compare the effects of thickness vs. hardness in scratch test measurements.

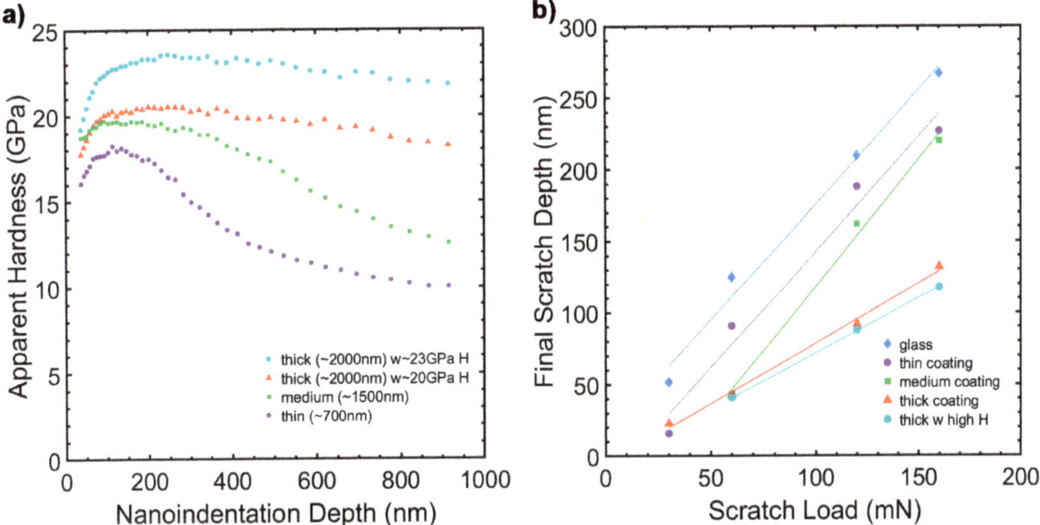

Figure 3. (a) Measured nanoindentation hardness vs. depth for a series of single-layer SiO_xN_y films of varying thickness coated on aluminosilicate glass substrates. (**b**) Results of Berkovich diamond tip nano-scratch test on the same single-layer SiO_xN_y films whose hardness profiles are shown in Figure 3a. Both of the 2000 nm thick films create a >50% scratch depth reduction relative to uncoated chemically strengthened aluminosilicate glass.

Figure 3b shows the results of Berkovich diamond nano-scratch testing on the same series of SiO_xN_y films. The nano-scratch depths cover a range up to 270 nm deep, which falls within the range that has been observed for real world micro-ductile scratches in consumer electronics devices, as described above. Interestingly, the 2000 nm thick coatings were the only ones that maintained their ability to significantly reduce scratch depth over this entire range of scratch severity. These thicker coatings were found to reduce the scratch depths by >50% compared to uncoated chemically strengthened aluminosilicate glass. A modest increase in coating hardness from 20 GPa to 23 GPa for the 2000 nm thick films resulted in only a marginal benefit, suggesting that coating thickness (hardness vs. depth profile) had a much stronger influence on scratch resistance over the ranges tested. These results suggest that most common AR coatings, which typically have total coating thicknesses of about 300–500 nm (including both high- and low-index materials) should not be expected to significantly reduce scratch severity in this range of depth and force when compared to chemically strengthened glass. This is true even if those common AR coatings incorporate layers of high-hardness materials.

3.3. Mechanically Tunable Multilayer Anti-Reflective Hardcoatings (AR-HCs)

To extend the analysis above to multilayer anti-reflection coatings, a series of anti-reflective hardcoatings (AR-HCs) were fabricated with nearly identical optics, but varying hardness vs. depth profiles. The fabricated coatings are depicted schematically in Figure 4, with detailed layer thicknesses given in the inset table. These coatings use the modular opto-mechanical design approach of [16]. An anti-reflective series of layers, also called impedance-matching layers, between the glass substrate and a central hardcoating layer substantially de-couples the optical performance of the multilayer stack from the thickness of this central layer, which is a unique feature not present in previously common anti-reflection coating designs. This allows the mechanical tuning of these anti-reflection coatings through varying the thickness of this central hardcoating layer, enabling the creation of varying hardness vs. depth profiles while maintaining a nearly identical optical performance. Upon evaluation of the hardness and scratch resistance of these anti-reflection coatings, optimal designs can be chosen that have the required scratch resistance for specific applications.

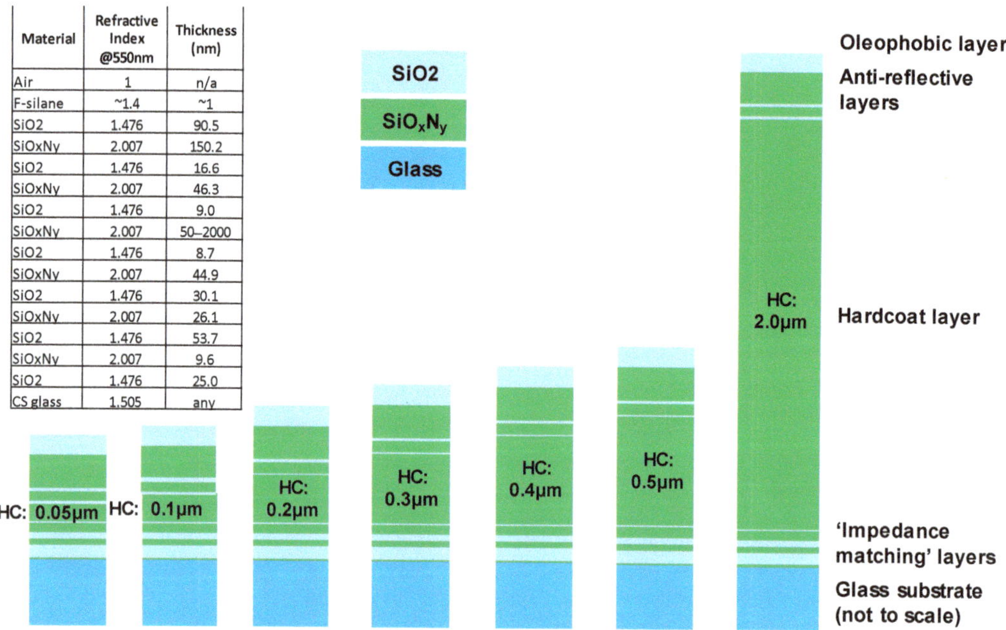

Material	Refractive Index @550nm	Thickness (nm)
Air	1	n/a
F-silane	~1.4	~1
SiO2	1.476	90.5
SiOxNy	2.007	150.2
SiO2	1.476	16.6
SiOxNy	2.007	46.3
SiO2	1.476	9.0
SiOxNy	2.007	50–2000
SiO2	1.476	8.7
SiOxNy	2.007	44.9
SiO2	1.476	30.1
SiOxNy	2.007	26.1
SiO2	1.476	53.7
SiOxNy	2.007	9.6
SiO2	1.476	25.0
CS glass	1.505	any

Figure 4. Schematic illustration of fabricated mechanically tunable anti-reflection coatings. Detailed layer design given in inset table.

Figure 5 illustrates the measured first-surface reflectance of the series of AR-HCs compared to an uncoated chemically strengthened (CS) glass control sample. Of note is the strikingly similar reflectance vs. wavelength performance shown for all coatings, despite the variation in central hardcoating layer thickness. This is a unique feature enabled by the modular anti-reflective hardcoating design approach, and enables visual or optical comparisons of damage resistance that are not confounded with changes in AR coating optics.

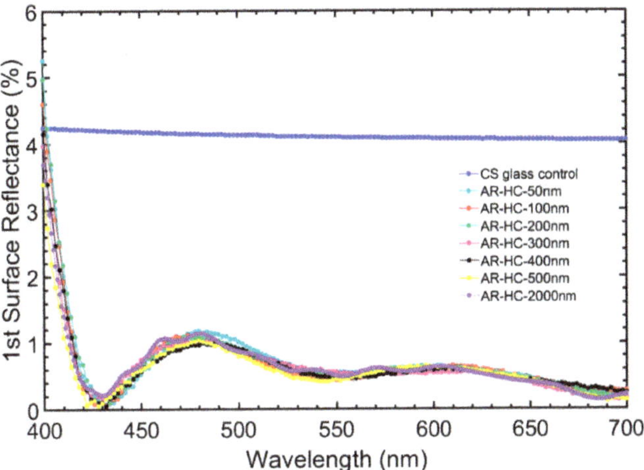

Figure 5. First-surface reflectance for series of anti-reflection coating designs depicted schematically in Figure 4, compared to chemically strengthened (CS) glass control sample. The coatings have remarkably similar optical performance, despite their quasi-continuously varying central hardcoat layer thickness. This illustrates the unique level of optical invariance that can be achieved with this design approach.

Figure 6 shows the measured nanoindentation depth vs. hardness profiles for the series of AR-HCs fabricated for this study, compared to uncoated chemically strengthened (CS) glass. As was seen for the single-layer hardcoatings, the hardness vs. depth profiles of the AR-HCs are strongly dependent on the thickness and amount of hardcoating material included in the multilayer design. Based on the single-layer results illustrated above, it is reasonable to expect that this series of AR-HCs would also exhibit strong variations in their scratch resistance performance.

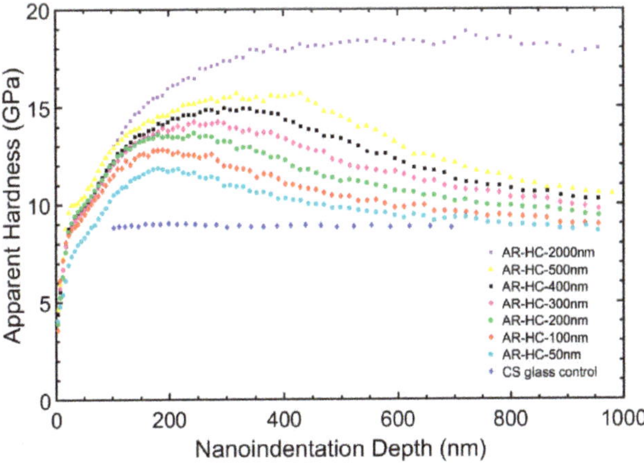

Figure 6. Measured nanoindentation hardness vs. depth profiles for series of anti-reflective hardcoatings compared to chemically strengthened (CS) glass control sample. The thickest anti-reflective hardcoatings (AR-HC) maintains high hardness to depths of 1000 nm and beyond, while the thinner AR-HCs have high hardness over a narrower range of depths, and begin to approach the hardness of the CS glass substrate at varying depths from about 500 nm and greater, depending on their thickness.

3.3.1. Nanoindentation Hardness Profile Modeling of AR-HCs

In order to further understand the fundamentals of hardness profile generation in these complex multilayer AR-HC systems, the nanoindentation process of coated glass was modeled using the finite element method (FEM) in the commercial software package ABAQUS v. 2018. An axisymmetric model was used to reduce the computational time and a conical indenter tip with a semi-angle of 70.3° was assumed, which generates the same contact area-to-depth ratio as the Berkovich tip. In practice, a sharp Berkovich indenter with a zero-tip radius does not exist. The radius of a new indenter tip is in the range of 40 nm–50 nm, which can increase to 100~200 nm or more as it wears. In this study, our indenter was assumed to have a 100 nm tip bluntness, which was incorporated in the model. The height and width of the FEM model were selected to be 600 and 200 μm, respectively, to avoid any sample size effect on the nanoindentation load–depth curve. Adaptive meshing was used with very fine mesh in the region directly underneath the indenter, with a smallest element size of 5 nm. Mesh convergence studies ensured proper mesh refinement had been achieved. A typical FEM model setup is shown in Figure 7.

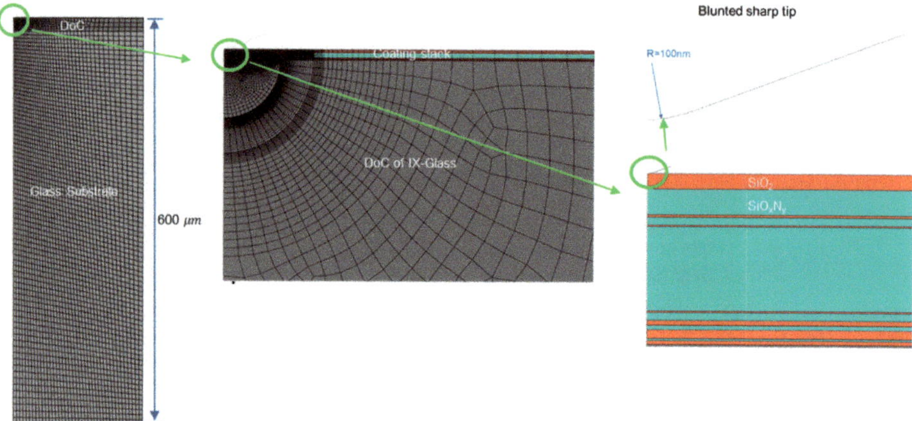

Figure 7. Example FEM model used to simulate the nanoindentation of AR-HC coating stacks on chemically strengthened glass substrates. The region of compressive stress in the ion-exchanged (IX) glass is denoted by the depth of the compressive layer (DoC).

In this modeling analysis, all the materials were assumed to be homogenous, isotropic, elastic perfectly-plastic, with the adoption of von Mises yield criterion [27,28]. Displacement history from the experiments were used as input for the FEM simulations. The Poisson's ratio was assumed to be 0.22, and the Young's modulus values, determined by experimentation, were used as input for the FEM simulations. Nanoindentation load as a function of prescribed displacement is the output of a simulation. To obtain the material properties of individual high and low-index material films, a single layer coating of each material on glass is first fabricated and measured. A classical inverse finite element approach was used to identify the material properties of the coating and glass substrate. This first step involves fitting the model to experimental data. Next, predictive modeling was carried out on complex multilayer designs, using these measured single-layer properties as inputs. These single-layer input properties used the FEM model are summarized in Table 1 below.

Table 1. Material properties used as inputs to FEM numerical study.

Material	Young's Modulus (GPa)	Poisson's Ratio	Yield Stress (GPa)
SiO_xN_y	230	0.22	12.4
SiO_2	70	0.22	4.0
Glass (CS layer)	78	0.22	4.8
Glass (bulk)	72	0.22	4.5

To extract hardness as a function of nanoindentation depth, continuous stiffness measurement (CSM) nanoindentation must be simulated. To this end, the nanoindenter tip was given a small amplitude of vibration during the loading stage. For our simulation, displacement history of the tip was prescribed by a user-defined "Amplitude" curve in ABAQUS to impose a very small (~1 nm or less) harmonic unloading. The maximum time increment was limited in such a way that history output can have a high sampling rate to capture all these 1 nm "unloading" portions during the overall loading stage.

A parametric study was carried out to investigate the effect of the H-layer (defined as the central, varying thickness SiO_xN_y layer in the AR-HC stacks) thickness. The load vs. nanoindentation depth curves were extracted from numerical simulations, as shown in Figure 8a. The hardness responses were then calculated using Equations (1)–(5), following the Oliver–Pharr method [21,22]. As can be seen in Figure 8b, a thicker H-layer can significantly enhance the hardness response.

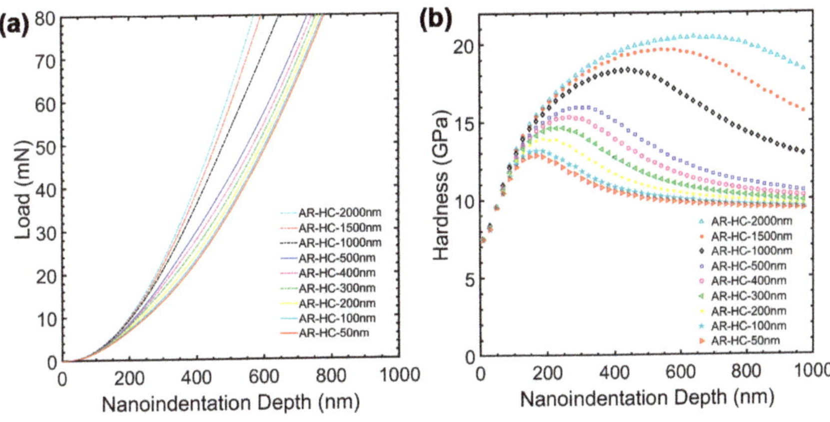

Figure 8. Modeling prediction of 13-layer AR-HC stack on glass indented by Berkovich tip with 100 nm bluntness (**a**) load vs. depth (**b**) hardness vs. depth. H-layer thickness ranges from 50 nm to 2000 nm.

Modeling predictions are compared with experimental data for various stacks, as shown in Figure 9. Overall, the modeling predictions show reasonably good agreement with the experimental data, especially for stacks with thicker H-layers. When the H-layer is only 50 nm (Figure 9g), there is some noticeable difference between the experiment and modeling outcomes. One of the possible reasons for this discrepancy is that the same set of material properties were used for all the numerical simulations. Those material properties were calibrated based on measurements of a single layer coating that has a thickness more than 1 μm. It is likely that a 50 nm H-layer might have different mechanical properties, due to the development of coating microstructure and density during film deposition, as compared to a 2000 nm H-layer. Nevertheless, this analysis shows the capability of FEM to predictively model hardness vs. depth profiles for complex multilayer (here, 13-layer) stacks, starting from single-layer measured experimental data used to establish the input material properties in the FEM model. This can significantly improve the coating stack design exploration and can be readily paired with optical modeling.

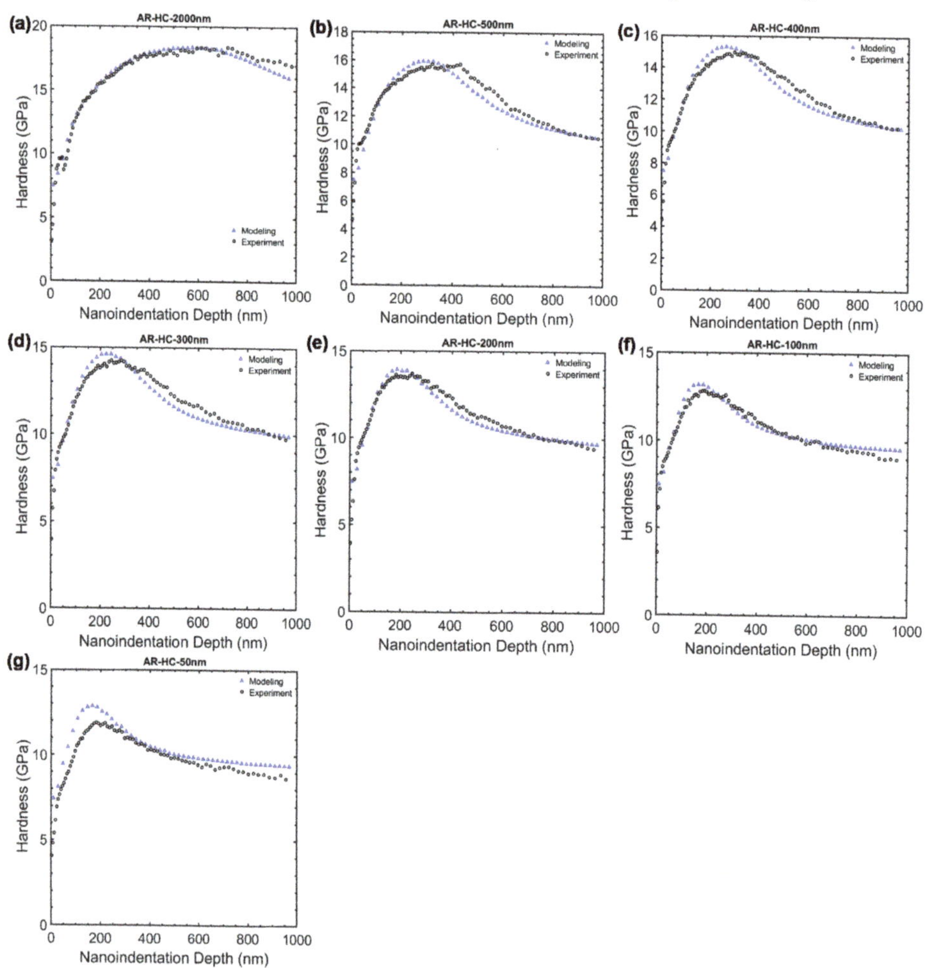

Figure 9. Nanoindentation hardness vs. depth responses of 13-layer AR-HC stacks on glass indented by Berkovich tip with 100 nm bluntness (**a**) 2000 nm thick H-layer (**b**) 500 nm thick H-layer (**c**) 400 nm thick H-layer (**d**) 300 nm thick H-layer (**e**) 200 nm thick H-layer (**f**) 100 nm thick H-layer (**g**) 50 nm thick H-layer. FEM predictive modeling shows generally good agreement with experimental results, as shown by the comparison of blue (modeling) and black (experimental) curves.

3.3.2. Scratch Resistance Testing of AR-HCs

In order to evaluate the practical scratch resistance of the fabricated AR-HCs and compare their hardness profiles to their resistance to simulated real-world surface damage, the previously described Al_2O_3 abrasion and garnet scratch tests were employed. Both of these tests were found to generate scratches with depths comparable to those found in our field studies, with the garnet scratch test creating damage having the highest similarity to the scratch events observed from the field, due to the single-pass nature of both the garnet scratch test and the observed field scratches. Figure 10 shows optical scattered light images of the AR-HC samples compared to control chemically strengthened glass. From the optical images, it is apparent that the surface durability improves with the increasing hardcoat layer thickness, most notably and consistently for those coatings with an H-layer thickness above ~400 nm and greater (total coating thickness of ~850 nm and greater). This greater coating thickness correlates with hardness sustained at deeper depths as shown through the nanoindentation hardness results above. AR-HC-2000 nm shows particularly excellent durability in this test, with abrasion resistance dramatically exceeding that of the chemically strengthened glass (control sample with no hardcoating).

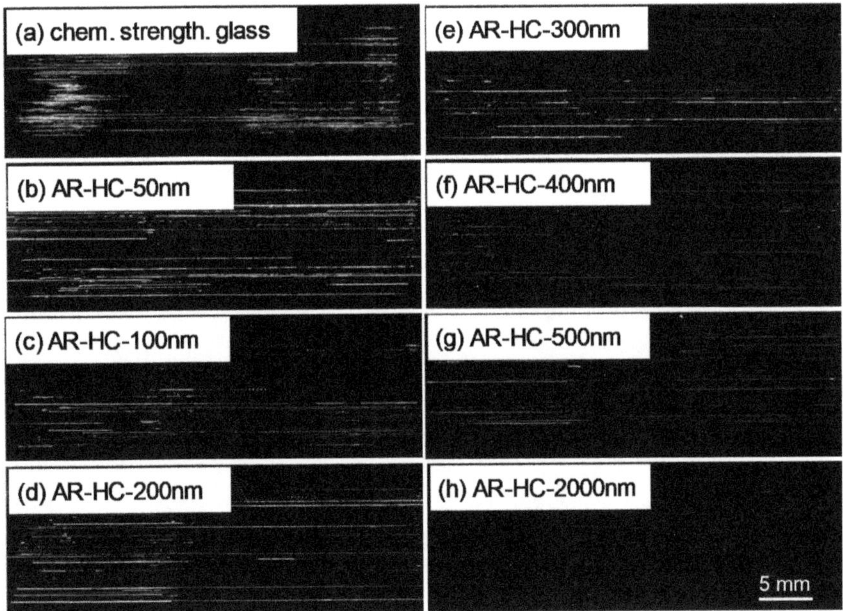

Figure 10. AR-HC samples abraded with 400 grit Al_2O_3 sandpaper for 50 cycles with 700 g applied load. Anti-reflective (AR) samples incorporating a hardcoating thickness of ~400 nm or greater clearly show improved abrasion resistance, while the performance is less consistent for samples having a lower thickness and a corresponding shallower depth of high hardness.

Abrasion resistance is further quantified in this test using a reflected scattered light measurement (specular component excluded, SCE, reflectance measurement). The results shown in Figure 11 indicate that this method (Al_2O_3 abrasion + SCE measurement) can be a useful means of quantification, as the trends of increasing abrasion resistance with hardcoat thickness are fairly consistent, correlating with the previously shown nanoindentation hardness vs. depth profiles. The SCE measurement appears to be more definitive when more severe damage is present, and may have more difficulty than the human eye or digital imaging methods in differentiating between samples with relatively lighter damage, such between the AR-HC-500 nm and AR-HC-2000 nm samples. While providing useful quantification of surface durability, it is important to note that the reciprocating abrasion method does not generate the same damage modes as those typically observed in real-world use for consumer electronics (which are single-event scratches).

Figure 12 shows optical scattered light images of the same AR-HC sample series after 1 kg garnet scratch testing. Due to the single-pass nature of this test, which approximates real-world scratch conditions, relatively fewer scratches were created, making the results more noisy and random. Nevertheless the trend of improved scratch resistance is still visible in this test, and the 2000 nm HC layer sample again shows excellent durability in this test, with scratch resistance dramatically exceeding that of the chemically strengthened glass control.

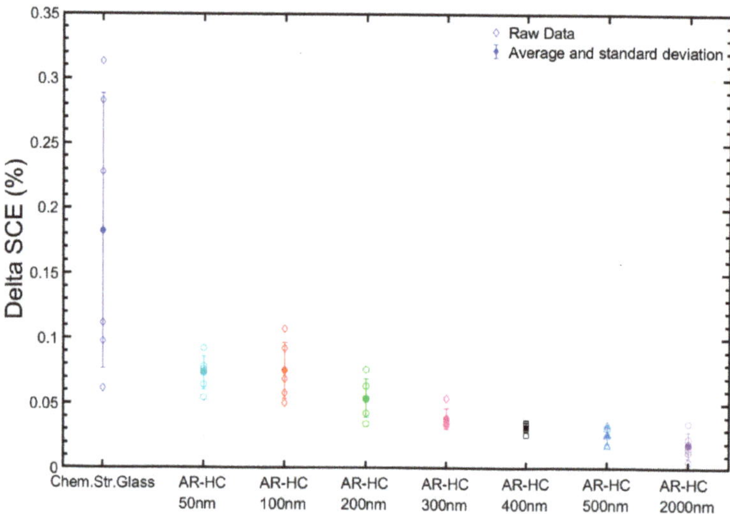

Figure 11. Reflected scattered light (SCE) measurement on Al_2O_3-abraded AR-HC samples. The trend of increasing abrasion resistance with hardcoat thickness is apparent under these conditions, with samples having H-layer $\geq$400 nm (total thickness $\geq$850 nm) performing particularly well in this test.

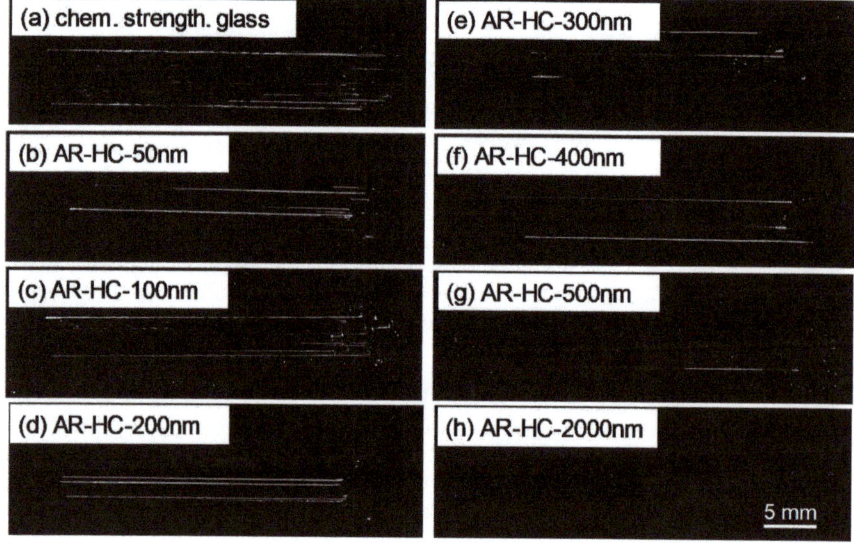

Figure 12. AR-HC samples scratched with 150 grit garnet sandpaper for 1 stroke with 1 kg applied load. AR samples incorporating a hardcoating thickness of ~500 nm or greater clearly show improved scratch resistance, while the performance is less consistent for samples having a lower thickness and a corresponding shallower depth of high hardness.

A scattered light quantification method would be noisy and less representative when paired with the single-pass garnet scratch test, due to the relatively random and low number of scratches generated in this test. As such, the garnet scratch severity was quantified through AFM scratch depth measurements for each scratch visible to the naked eye, in the center of the scratch path. Results are shown in Figure 13, giving an indication of both scratch frequency (number of visible scratches) and scratch severity (depth). Of note, the most frequent scratches are in the depth range of those most commonly observed in the field analysis summarized in Figure 1, while a few scratches of greater depth indicate the severity of this test method. The AR-HC samples having an H-layer thickness of 500 nm and greater (total coating thickness of 950 nm and greater) show the clearest reduction in scratch frequency and depth in this test. The AR-HC-2000 nm sample consistently yields zero visible scratches in this test.

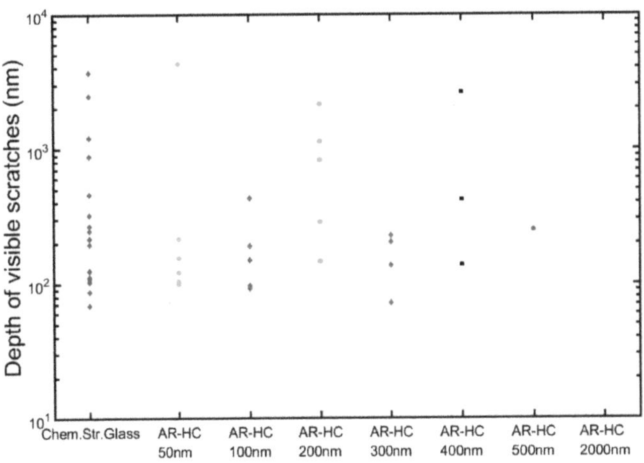

Figure 13. Depth of scratches visible to the unaided human eye after 1 kg garnet scratch testing, as measured using atomic force microscopy (AFM). AR-HC samples with greater thickness and a corresponding greater depth of high hardness show a reduced frequency of scratches, particularly for samples with H-layer thickness ≥500 nm (total thickness ≥950 nm). Sample AR-HC-2000 nm consistently yields no visible scratches in this test, thus, there are no data points for that sample.

4. Discussion

Resistance to micro-ductile deformation/scratching is directly associated with the hardness of the material; higher hardness imparts greater resistance to this type of damage. When using hard coatings to impart scratch resistance, the thickness of the coating (and the resulting hardness vs. depth profile) plays a critical role in deriving the full benefit of the coating. If a coating is too thin, its hardness will not be maintained over a sufficient depth range to prevent moderate to severe scratches in real world scenarios. The contact mechanics responsible for these hardness vs. depth profile effects have been known for some time, but the implications for practical scratch resistant coatings do not appear to have been widely recognized or practiced until very recently [16]. Many attempts to utilize scratch resistant coatings have failed because the critical role of coating thickness has not been fully appreciated—the temptation is to utilize high hardness films that are no more than ~100–200 nanometers thick. The required hardness and thickness levels are intimately related to the nature of the scratching that one is trying to prevent, i.e., the severity of the real-world scratch events that will be encountered in a particular application. Based on analysis and understanding gained from the touch-screen consumer device industry, combined with modeling and experimental analysis of anti-reflection coatings with tunable hardness vs. depth profiles, the types of hardness profiles that are necessary and effective to dramatically reduce moderate to severe scratching in these types of applications are defined in the present work. In particular, coatings with a hardness above ~15 GPa that is maintained over a depth range of ~200–800 nm have been shown to be particularly effective in preventing real-world scratches. The three distinct scratch and abrasion tests utilized here clearly indicate that coatings with depth-of-hardness profiles falling well below these levels are more prone to scratching, and have scratch resistance levels that are close to, or even worse than, chemically strengthened glass. One important factor to keep in mind for anti-reflective coatings is that a scratch with a 50–500 nm depth in an AR coating will typically be much more visible than a scratch of the same depth on uncoated glass, due to the optical interference nature of the coatings which drives high optical contrast between the scratched and unscratched regions. Thus, it is particularly important to minimize scratch frequency in these AR coatings. In order to maintain high hardness over the indicated depth range and nearly eliminate scratching under simulated real-world test conditions, a total coating thickness of ~2.5 microns (with ~2 microns thickness of high-hardness material) has been found to be particularly excellent. The scratch resistance levels for these coatings, exemplified by AR-HC-2000 nm, dramatically exceed those of chemically strengthened glass. For applications with less stringent scratch or damage requirements, anti-reflective hardcoating thicknesses of around ~1 micron (with ~0.5 microns of high-hardness material) have been found to be effective at reducing overall scratch visibility to a level that is moderately better than chemically strengthened glass (which itself is far more scratch resistant than other materials such as polymers).

In addition to the previously described modular opto-mechanical design approach, which enables the fabrication of mechanically tunable optical coatings with nearly invariable optical properties, the utility of additional design and analysis tools is demonstrated here. FEM modeling of hardness was found to adequately predict the hardness vs. depth profiles of multilayer AR-HCs, and this approach can be coupled with optical modeling for multifunctional coating design. Various scratch test methods were evaluated and compared to the trends seen in hardness vs. depth profiles. All of the scratch tests were chosen because they generate scratches of a similar depth and type as those observed in field analysis. These scratch test methods showed a clear trend of agreement with the hardness vs. depth profile analysis—samples with the highest hardness maintained at depths of 500 nm and greater showed especially excellent durability in aggressive scratch and abrasion testing. The garnet scratch test shown here produces scratches that have the greatest similarity to real-world scratches observed in our consumer electronics field trials, however, the single-pass scratch event produces a high level of randomness that makes it more difficult to quantitatively measure the light scattering from these scratches (as compared to the Al_2O_3 reciprocating abrasion method). Thus, the Al_2O_3 abrasion method is also useful for quantifying surface damage resistance, due to the ease of quantification through light scattering illustrated here. The diamond nano-scratch method employed is highly precise and useful for developing fundamental understanding, but is less useful for establishing practical real-world durability requirements, due to the fact that diamond is an uncommon abradant in the real world.

5. Conclusions

Mechanically tunable anti-reflective hardcoatings were fabricated, demonstrating the ability to quasi-continuously vary the thickness of a central hardcoat layer in the multilayer stack design, with minimal impact on the anti-reflective optical performance. The fundamentals of hardcoat layer thickness vs. nanoindentation hardness depth profiles were explored through FEM modeling and experimentation. Nanoindentation hardness depth profiles were correlated with practical scratch resistance using three different scratch and abrasion test methods, simulating real world scratch events as observed from consumer electronics field studies. For high practical scratch resistance, coating designs with a hardness of >15 GPa maintained over depths of 200–800 nm, were found to be particularly excellent. The tools and design approaches illustrated here can be extended to create mechanically durable multilayer optical interference coatings of various kinds for multiple applications, including wavelength-selective optics [29], bandpass filters, ophthalmic lenses, IR sensors, and more.

Author Contributions: Conceptualization, K.W.K., L.L., J.J.P., C.A.P., A.M., C.K.W., and S.D.H.; methodology, T.X., B.Z., K.W.K., L.L., J.J.P., E.L.N., K.B.R., and S.D.H.; validation, E.L.N., C.-G.K., D.-G.M., S.-Y.O., J.-K.O., B.Z., K.B.R., and J.J.P.; formal analysis, T.X., K.W.K., L.L., S.D.H., and C.-G.K., investigation, C.-G.K., D.-G.M., S.-Y.O., J.-K.O., C.A.P., T.X., B.Z., and J.J.P.; writing—original draft preparation, J.J.P., T.X., and S.D.H.; writing—review and editing, T.X., S.D.H., and C.K.W.; visualization, T.X., J.J.P., B.Z., and S.D.H.; supervision, J.-H.O., A.M., and C.K.W.; project administration, J.-H.O., A.M., and C.K.W. All authors have read and agreed to the published version of the manuscript.

Funding: This research received no external funding.

Institutional Review Board Statement: Not applicable.

Informed Consent Statement: Not applicable.

Data Availability Statement: The data presented in this study are available on request from the corresponding author.

Acknowledgments: The authors would like to acknowledge Loretta Moses, Jaymin Amin, Jason Harris, Ananth Subramanian, Anthony Furstoss, Leon Reed, Marina Belyustina, Guangli Hu, Heather Decker, Nicholas Walker, Sanjay Sinha, Robert Bellman, John Langstrand, Junghun Yun, Jinah Yoo, Heeseon You, Jiman Lee, Eunhae Ji, Bill McKendrick, Hazel Russell, Arnette Brooks, Brandy Fuller, Phyllis Markowski, Josh Jacobs, Jum Kim, Robert Lee, and Odessa Petzold, for their assistance in experiments, analysis, logistical support, and moral support.

Conflicts of Interest: All authors are employees of Corning Incorporated, which may choose to market or sell technical products related to the subject matter of this article.

References

1. Macleod, H.A. *Thin-Film Optical Filters*, 5th ed.; CRC Press: Boca Raton, FL, USA, 2017.
2. Baumeister, P.W. Methods of altering the characteristics of a multilayer stack. *J. Opt. Soc. Am. A* **1962**, *52*, 1149–1152. [CrossRef]
3. Southwell, W.H. Coating design using very thin high-and low-index layers. *Appl. Opt.* **1985**, *24*, 457–460. [CrossRef]
4. Raut, H.K.; Ganesh, V.A.; Nair, A.S.; Ramakrishna, S. Antireflective coatings: A critical, in-depth review. *Energy Environ. Sci.* **2011**, *4*, 3779–3804. [CrossRef]
5. Aiken, D.J. High performance anti-reflection coatings for broadband multi-junction solar cells. *Sol. Energy Mater. Sol. Cells* **2000**, *64*, 393–404. [CrossRef]
6. Fink, Y.; Winn, J.N.; Fan, S.; Chen, C.; Michel, J.; Joannopoulos, J.D.; Thomas, E.L. A dielectric omnidirectional reflector. *Science* **1998**, *282*, 1679–1682. [CrossRef] [PubMed]
7. Szipöcs, R.; Spielmann, C.; Krausz, F.; Ferencz, K. Chirped multilayer coatings for broadband dispersion control in femtosecond lasers. *Opt. Lett.* **1994**, *19*, 201–203. [CrossRef]
8. Weber, M.F.; Stover, C.A.; Gilbert, L.R.; Nevitt, T.J.; Ouderkirk, A.J. Giant birefringent optics in multilayer polymer mirrors. *Science* **2000**, *287*, 2451–2456. [CrossRef]
9. Beauchamp, W.T.; Tuttle-Hart, T. UV/IR Reflecting Solar Cell Cover. U.S. Patent 5,449,413, 9 December 1995.
10. Lequime, M. Tunable thin film filters: Review and perspectives. In *Proceedings of the SPIE-Optical Systems Design, St. Etienne, France, 30 September 2003*; Advances in Optical Thin Films; International Society for Optics and Photonics: Bellingham, WA, USA, 2004; Volume 5250, pp. 302–311.
11. Williams, C.; Hong, N.; Julian, M.; Borg, S.; Kim, H.J. Tunable mid-wave infrared Fabry-Perot bandpass filters using phase-change GeSbTe. *Opt. Express* **2020**, *28*, 10583–10594. [CrossRef]
12. Scobey, M.A.; Stupik, P.D. Stable ultranarrow bandpass filters. In *Proceedings of the SPIE's International Symposium on Optics, Imaging, and Instrumentation, San Diego, CA, USA, 7 September 1994*; Optical Thin Films IV: New Developments; International Society for Optics and Photonics: Bellingham, WA, USA, 1994; Volume 2262, pp. 37–46.
13. Serényi, M.; Rácz, M.; Lohner, T. Refractive index of sputtered silicon oxynitride layers for antireflection coating. *Vacuum* **2001**, *61*, 245–249. [CrossRef]
14. Wang, Y.; Cheng, X.; Lin, Z.; Zhang, C.; Zhang, F. Optimization of PECVD silicon oxynitride films for anti-reflection coating. *Vacuum* **2003**, *72*, 345–349. [CrossRef]
15. Hart, S.D.; Koch, K.W.; Paulson, C.A.; Price, J.J. Durable and Scratch-Resistant Anti-Reflective Articles. U.S. Patent 9,335,444, 10 May 2016.
16. Paulson, C.A.; Price, J.J.; Koch, K.W.; Kim, C.-G.; Oh, J.-H.; Lin, L.; Subramanian, A.N.; Zhang, B.; Amin, J.; Mayolet, A.; et al. Industrial-grade anti-reflection coatings with extreme scratch resistance. *Opt. Lett.* **2019**, *44*, 5977–5980. [CrossRef] [PubMed]
17. Wang, J.; Bouchard, J.P.; Hart, G.A.; Oudard, J.F.; Paulson, C.A.; Sachenik, P.A.; Price, J.J. Silicon oxynitride based scratch resistant antireflective coatings. In *Proceedings of the SPIE Defense + Security, Orlando, FL, USA, 8 May 2018*; Advanced Optics for Defense Applications: UV through LWIR III; International Society for Optics and Photonics: Bellingham, WA, USA, 2018; Volume 10627, p. 106270G.
18. Paulson, C.A. Scratch-Resistant and Optically Transparent Materials and Articles. U.S. Patent 10,603,870, 31 March 2020.
19. Musil, J.; Baroch, P.; Vlček, J.; Nam, K.H.; Han, J.G. Reactive magnetron sputtering of thin films: Present status and trends. *Thin Solid Films* **2005**, *475*, 208–218. [CrossRef]

20. Rademacher, D.; Zickenrott, T.; Vergöhl, M. Sputtering of dielectric single layers by metallic mode reactive sputtering and conventional reactive sputtering from cylindrical cathodes in a sputter-up configuration. *Thin Solid Films* **2013**, *532*, 98–105. [CrossRef]
21. Oliver, W.C.; Pharr, G.M. An improved technique for determining hardness and elastic modulus using load and displacement sensing indentation experiments. *J. Mater. Res.* **1992**, *7*, 1564–1583. [CrossRef]
22. Oliver, W.C.; Pharr, G.M. Measurement of hardness and elastic modulus by instrumented indentation: Advances in understanding and refinements to methodology. *J. Mater. Res.* **2004**, *19*, 3–20. [CrossRef]
23. Lucas, B.N.; Oliver, W.C.; Swindeman, J.E. The dynamics of depth-sensing, frequency-specific indentation testing. *Mat. Res. Soc. Symp. Proc.* **1998**, *3*, 14. [CrossRef]
24. Irland, M.J.; Schermer, E.B. A method for evaluating the abrasion resistance of optical surfaces and thin films. *Appl. Opt.* **1964**, *3*, 751–754. [CrossRef]
25. Evans, D.; Zuber, K.; Hall, C.; Griesser, H.J.; Murphy, P. Abrasion resistance of thin film coatings as measured by diffuse optical scattering. *Surf. Coat. Technol.* **2011**, *206*, 312–317. [CrossRef]
26. Schneider, J.; Schula, S.; Weinhold, W.P. Characterisation of the scratch resistance of annealed and tempered architectural glass. *Thin Solid Films* **2012**, *520*, 4190–4198. [CrossRef]
27. Sun, Y.; Bloyce, A.; Bell, T. Finite element analysis of plastic deformation of various TiN coating/substrate systems under normal contact with a rigid sphere. *Thin Solid Films* **1995**, *271*, 122–131. [CrossRef]
28. Care, G.; Fischer-Cripps, A.C. Elastic-plastic indentation stress fields using the finite-element method. *J. Mater. Sci.* **1997**, *32*, 5653–5659. [CrossRef]
29. Koch, K.W.; Lin, L.; Price, J.J.; Kim, C.-G.; Moon, D.-G.; Oh, S.-Y.; Oh, J.-K.; Oh, J.-H.; Paulson, C.A.; Zhang, B.; et al. Wavelength-Selective Coatings on Glass with High Hardness and Damage Resistance. *Coatings* **2020**, *10*, 1247. [CrossRef]

Article

Influence to Hardness of Alternating Sequence of Atomic Layer Deposited Harder Alumina and Softer Tantala Nanolaminates

Helle-Mai Piirsoo *, Taivo Jõgiaas, Peeter Ritslaid, Kaupo Kukli and Aile Tamm

Institute of Physics, University of Tartu, W. Ostwaldi Str. 1, 50411 Tartu, Estonia; taivo.jogiaas@ut.ee (T.J.); peeter.ritslaid@ut.ee (P.R.); kaupo.kukli@ut.ee (K.K.); aile.tamm@ut.ee (A.T.)
* Correspondence: helle-mai.piirsoo@ut.ee

Abstract: Atomic layer deposited amorphous 70 nm thick Al_2O_3-Ta_2O_5 double- and triple-layered films were investigated with the nanoindentation method. The sequence of the oxides from surface to substrate along with the layer thickness had an influence on the hardness causing rises and declines in hardness along the depth yet did not affect the elastic modulus. Hardness varied from 8 to 11 GPa for the laminates having higher dependence on the structure near the surface than at higher depths. Triple-layered Al_2O_3/Ta_2O_5/Al_2O_3 laminate possessed the most even rise of hardness along the depth and possessed the highest hardness out of the laminates (11 GPa at 40 nm). Elastic modulus had steady values along the depth of the films between 145 and 155 GPa.

Keywords: nanoindentation; nanolaminates; atomic layer deposition; nanoengineering

Citation: Piirsoo, H.-M.; Jõgiaas, T.; Ritslaid, P.; Kukli, K.; Tamm, A. Influence to Hardness of Alternating Sequence of Atomic Layer Deposited Harder Alumina and Softer Tantala Nanolaminates. *Coatings* **2022**, *12*, 404. https://doi.org/10.3390/coatings12030404

Academic Editors: Rongguang Xu, Zhitong Chen, Peijian Chen and Guangjian Peng

Received: 22 February 2022
Accepted: 17 March 2022
Published: 18 March 2022

Publisher's Note: MDPI stays neutral with regard to jurisdictional claims in published maps and institutional affiliations.

1. Introduction

Nanolaminates are being investigated for the possibility of nanoengineering, i.e., constructing thin films with a combination of multiple physical properties exactly suitable for a given application. Customizable properties dependent on the layered structure range from mechanical hardness [1–8] to magnetic susceptibility [2,9–11], and refraction properties [1].

The structure of layers constituting nanolaminates influences the mechanical properties of the film. For instance, different order of ZrO_2 and SnO_2 layers from surface to substrate in only 30 nm thick bilayer thin film caused a change of about 5 GPa in hardness and 100 GPa in the modulus [2]. For TiN/TiBN nanolayered films, it was found, that the layer thickness influenced film adhesion, wear resistance, and frictional coefficient in addition to hardness and elastic modulus [3]. The number of layers in ZnO/graphene nanolaminate influenced the hardness by ±6 GPa and elastic modulus by ±100 GPa [4]. ZrN layer thickness in reactive magnetron sputtered ZrN/SiNx nanolaminates had a correlation with the intrinsic growth stress [5]. The layered structure of ALD Al_2O_3/TiO_2 films also has had an influence on the residual stress in the film [6,7], as well as their hardness and elastic modulus [8].

Atomic layer deposition (ALD) of Al_2O_3 with $Al(CH_3)_3$ and H_2O precursors is a widely studied and known process [12,13]. Ta_2O_5 combined with Al_2O_3 in laminate thin films have been proposed for various applications due to their promising electrical properties, often dependent on the layered structure. Amorphous Al_2O_3-Ta_2O_5 quadruple-layered laminate showed higher dielectric strength compared to bilayers and laminates with eight layers [14]. The operation of high voltage metal-insulator-metal capacitors with Al_2O_3-Ta_2O_5 bilayers depended on the Al_2O_3/Ta_2O_5 thickness ratio [15]. Al_2O_3-Ta_2O_5 laminates have also exhibited resistive switching characteristics [16] and good corrosion resistance [17]. ALD of Ta_2O_5 with $Ta(OEt)_5$ and H_2O precursors has been developed before [18–22]. Deposition of Al_2O_3 or Ta_2O_5 at 300 °C results in amorphous oxides for both of the aforementioned ALD processes. The mechanical hardness of amorphous atomic layer deposited Al_2O_3 thin films lays between 7 and 12 GPa. It has been reported to be 10.5 [23] and 12 [24] GPa on Si substrates, 11 GPa on TiN/Si [25], 9.5 GPa on soda-lime glass [26], and 7.2 [27] and 11.8 [28] GPa on steel substrates. The elastic modulus of these Al_2O_3 films has varied from 68 (on glass) to 260 (on steel) GPa, while samples with Si substrate possessed intermediate values (145, 150, 220 GPa) [23–26,28]. The mechanical hardness of atomic layer deposited Ta_2O_5 film on a soda-lime glass substrate has been found to be 6.7 GPa, while the modulus was 96 GPa [26]. The hardness has been measured to be lower than that for Ta_2O_5 thin films fabricated with other methods—4.7 GPa for e-beam deposition on Co-Cr substrate [29] and 5.2 GPa for physical vapor deposition on Ti-6Al-4V substrate [30]. However, the elastic modulus was slightly higher in the films obtained with other methods compared to that in the films grown by ALD, i.e., 160 GPa with ion beam sputtering on Si [31] and 119 GPa with e-beam deposition on Co-Cr substrate [29].

In this work, double- and triple-layered amorphous Al_2O_3-Ta_2O_5 laminates with an overall thickness of about 70 nm were atomic layer deposited while changing the sequence of the layers from surface to substrate. Hardness and elastic modulus of the laminates were measured with nanoindentation, and the effects of the constituent layer thickness and deposition sequence on the mechanical properties of the laminates were analyzed.

2. Materials and Methods

2.1. Film Deposition

The layered films were produced by atomic layer deposition (ALD) in an in-house built reactor [32]. The metal precursors were $Al(CH_3)_3$ and $Ta(OC_2H_5)_5$, while H_2O was used as the oxygen source. Pulse times were 2/2/2/5 s for metal precursor, N_2 purge, water and the second purge, respectively. Evaporation temperature for $Ta(OC_2H_5)_5$ was $95 \pm 5\ ^\circ$C, and the growth temperature for both Al_2O_3 and Ta_2O_5 was $300 \pm 10\ ^\circ$C. Si (100) wafers were exploited as substrates. The number and sequence of deposition cycles are presented in Table 1. Reference Al_2O_3 and Ta_2O_5 films were deposited along with two double-layer laminates and two triple-layer laminates with different layering orders (Figure 1). An additional quadruple-layered laminate was also prepared.

Table 1. Denotations of films with the number of atomic layer deposition cycles with Al_2O_3 representing the $Al(CH_3)_3$ and water cycle and Ta_2O_5 representing the $Ta(OC_2H_5)_5$ and water cycle. Layer thicknesses according to X-ray reflectometry are presented in order from substrate to surface.

Denotation of Film	No. of ALD Cycles	Layer Thickness (nm)
Al_2O_3/Si	$600 \times Al_2O_3$	69 nm Al_2O_3
Ta_2O_5/Si	$800 \times Ta_2O_5$	61 nm Ta_2O_5
Ta_2O_5/Al_2O_3/Si	$340 \times Al_2O_3 + 470 \times Ta_2O_5$	38 nm Al_2O_3 + 35 nm Ta_2O_5
Al_2O_3/Ta_2O_5/Si	$470 \times Ta_2O_5 + 340 \times Al_2O_3$	34 nm Ta_2O_5 + 40 nm Al_2O_3
Al_2O_3/Ta_2O_5/Al_2O_3/Si	$227 \times Al_2O_3 + 313 \times Ta_2O_5 + 227 \times Al_2O_3$	29 nm Al_2O_3 + 23 nm Ta_2O_5 + 27 nm Al_2O_3
Ta_2O_5/Al_2O_3/Ta_2O_5/Si	$313 \times Ta_2O_5 + 227 \times Al_2O_3 + 313 \times Ta_2O_5$	23 nm Ta_2O_5 + 24 nm Al_2O_3 + 24 nm Ta_2O_5
Ta_2O_5/Al_2O_3/Ta_2O_5/Al_2O_3/Si	$170 \times Al_2O_3 + 268 \times Ta_2O_5 + 170 \times Al_2O_3 + 268 \times Ta_2O_5$	19 nm Al_2O_3 + 19 nm Ta_2O_5 + 19 nm Al_2O_3 + 20 nm Ta_2O_5

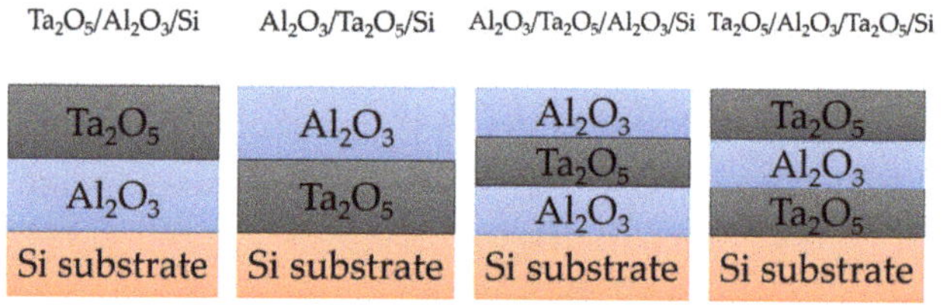

Figure 1. Schematics illustrating the layered structure of the laminates as the notation for each of the laminates is shown above.

2.2. Film Characterization

The elemental composition of the films was determined by wavelength dispersive X-ray fluorescence (WD-XRF) spectroscope ZSX-400 (Rigaku, Tokyo, Japan). The surface of the films was characterized by scanning electron microscopy (SEM) using Helios NanoLab 600 (FEI, Hillsboro, OR, USA). The phase composition of the films was determined by grazing incidence X-ray diffraction (GIXRD). The layered structure of the films was confirmed with the X-ray reflectivity (XRR) method. The XRD and XRR measurements were done with SmartLab™ (Rigaku, Tokyo, Japan) diffractometer, and for the analysis, AXES [33] software was used. In GIXRD measurements, grazing angle was 0.5°, scan step 0.04°, and scan rate 3°/min. Grazing angle varied till 7°, scan step and rate were 0.01° and 1.5°/min, respectively, for the XRR measurements.

2.3. Nanoindentation

Nanoindentation was carried out with Hysitron TriboIndenter TI980 (Bruker, Billerica, MA, USA) with a Berkovich tip. Peak load of 0.5 mN was applied in continuous stiffness measurement mode with tip frequency of 220 Hz, which obtains several tens of data points over a displacement range for a single indentation [34]. Loading, peak hold, and unloading times were set, respectively, to 35, 2, and 5 s. Fifteen indents in total were laterally separated by 10 μm. Gathered data were analyzed (linear interpolation, averaging) with Origin 2020 software. Prior to the measurements tip was calibrated on a fused quartz glass with a reduced modulus of 69.6 GPa and hardness of 9.25 GPa (Figure 2). The TriboIndenter was used in scanning probe microscopy mode to image the indents.

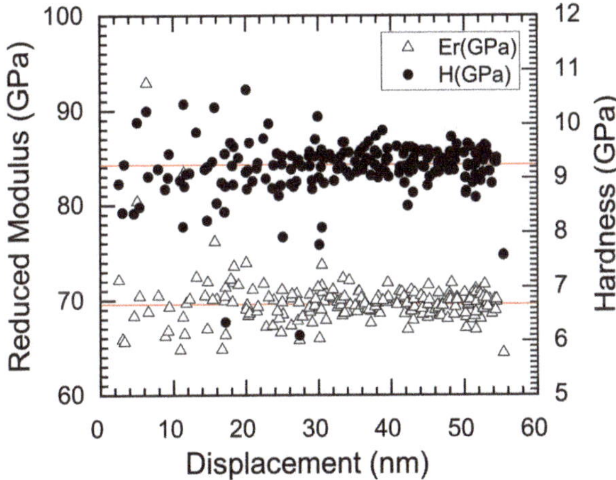

Figure 2. Calibration measurements for reduced modulus (▲) and hardness (●) along the depth of the reference quartz sample.

3. Results

3.1. Elemental Composition and Microstructure

The ALD growth rates were 0.11 nm/cycle for Al_2O_3 and 0.07 nm/cycle for Ta_2O_5 as determined with XRR and XRF methods and confirmed with ellipsometry (not shown) for reference samples. XRF results are presented in Table 2, revealing atomic ratios appreciably close to the stoichiometric Al_2O_3 and Ta_2O_5, as expected for the ALD processes [13,22].

SEM measurements revealed quite smooth, i.e., featureless surfaces for all the deposited films. An example surface is shown in Figure 3a. All the deposited films were X-ray amorphous, exemplified by the diffractogram of the laminates in Figure 3b. XRR measurement fittings confirmed the layered structures and thicknesses are presented in

Table 1. Average density of the oxides was 3.1 ± 0.2 g/cm^3 and 8.4 ± 0.2 g/cm^3 for Al$_2$O$_3$ and Ta$_2$O$_5$ layers, respectively.

Table 2. Elemental composition of the $\approx$70 nm thick films according to XRF.

Film	Al (at.%)	Ta (at.%)	O (at.%)
Al$_2$O$_3$/Si	41	-	59
Ta$_2$O$_5$/Si	-	25	75
Ta$_2$O$_5$/Al$_2$O$_3$/Si	21	13	66
Al$_2$O$_3$/Ta$_2$O$_5$/Si	20	12	68
Al$_2$O$_3$/Ta$_2$O$_5$/Al$_2$O$_3$/Si	27	8	65
Ta$_2$O$_5$/Al$_2$O$_3$/Ta$_2$O$_5$/Si	13	17	70
Ta$_2$O$_5$/Al$_2$O$_3$/Ta$_2$O$_5$/Al$_2$O$_3$/Si	19	14	67

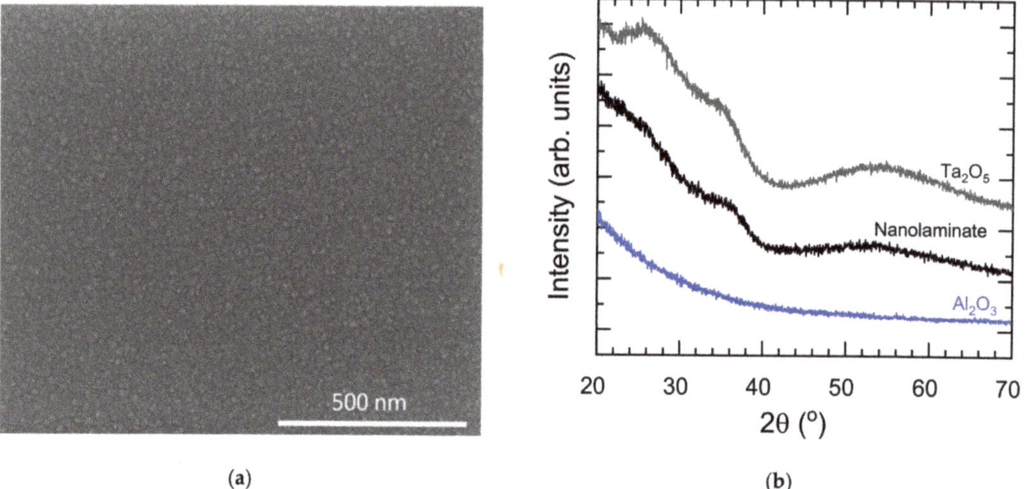

Figure 3. (a) SEM image of the surface of as-deposited Al$_2$O$_3$/Ta$_2$O$_5$/Al$_2$O$_3$/Si laminate, (b) GIXRD diffractograms of Al$_2$O$_3$-Ta$_2$O$_5$ nanolaminate and reference Ta$_2$O$_5$ and Al$_2$O$_3$ films.

3.2. Mechanical Properties of Amorphous Laminates

Figure 4 depicts the averaged hardness values with standard deviation for reference samples: Al$_2$O$_3$/Si, Ta$_2$O$_5$/Si, and the four-layered Ta$_2$O$_5$/Al$_2$O$_3$/Ta$_2$O$_5$/Al$_2$O$_3$/Si laminate films. The four-layered laminate is considered as a reference because it possessed a homogeneous hardness across displacements in the approximate range of the film thickness. The Si substrate possessed an average hardness of 13.5 ± 0.4 GPa and an average elastic modulus of 147 ± 3 GPa (not shown).

The hardness of Al$_2$O$_3$ had a steady incline with depth from around 12 GPa, after displacement of 11 nm, to 13 GPa at 40 nm. Similarly, the hardness of Ta$_2$O$_5$ film rose from 8 GPa at 11 nm to 10 GPa at 45 nm. The Al$_2$O$_3$ and Ta$_2$O$_5$ films exhibited a difference in hardness around 3–4 GPa over the measured displacement range. For the laminate, the thickness of layers is presented on the graph with dotted lines and arrows. The hardness of sample Ta$_2$O$_5$/Al$_2$O$_3$/Ta$_2$O$_5$/Al$_2$O$_3$/Si started from 9.5 GPa at 10 nm and then continued with a steady incline to 11 GPa with depth. The hardness-displacement graphs did not indicate the occurrence of indentation size effect, i.e., a rise in hardness near surface, which has been recorded for metal films in a similar displacement range [35].

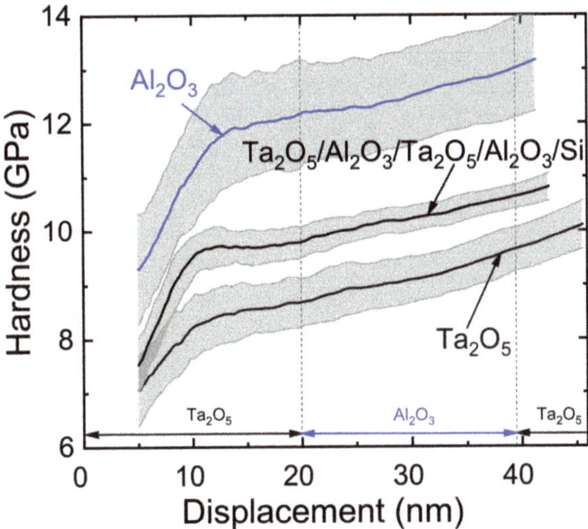

Figure 4. Averaged hardness of 15 indents with standard deviation shown as grey areas. Vertical dotted lines and arrows indicate the thickness and sequence of three topmost layers for the $Ta_2O_5/Al_2O_3/Ta_2O_5/Al_2O_3/Si$ laminate.

Figure 5a presents the hardness for the two double-layered laminates (Figure 1). For the $Al_2O_3/Ta_2O_5/Si$ film, the different mechanical hardness of individual layers is apparent decided on the basis of the wavelike shape of the curve as the hardness shows a maximum (10.5 GPa) at a displacement of 11 nm and a minimum (9.9 GPa) at 25 nm of depth. It should be noted that the interface between the layers was formed at 39 nm (Figure 5a).

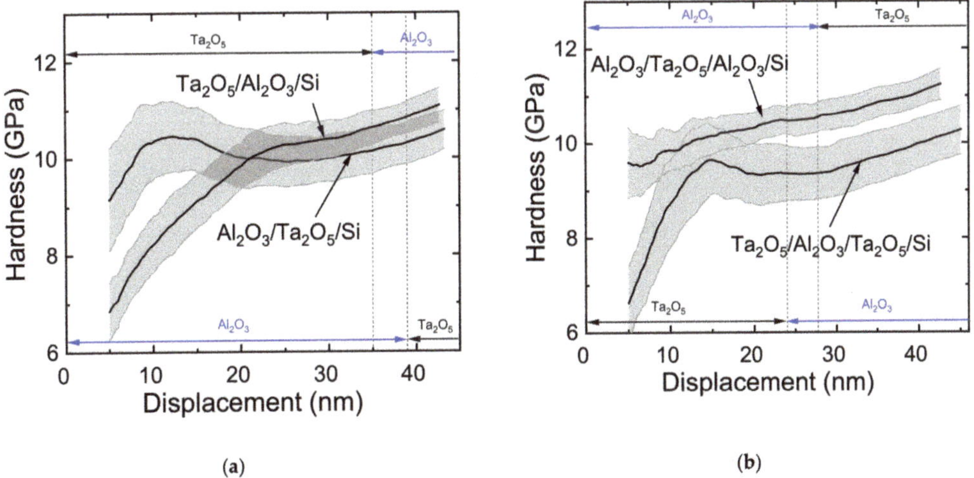

(a)   (b)

Figure 5. (**a**) Averaged hardness of 15 indents with standard deviation shown as grey areas. Vertical dotted lines and arrows indicate the thickness and sequence of layers for the double-layered laminates. (**b**) Averaged hardness of 15 indents with standard deviation shown as grey areas. Vertical dotted lines and arrows indicate the thickness and sequence of layers for the triple-layered laminates. Only two of the topmost layers are visualized on the graph.

For the $Ta_2O_5/Al_2O_3/Si$ laminate, the hardness depicted a steep rise until 20 nm of depth, however, the angle of incline changed twice, it decreased at 20 nm and slightly increased again at 33 nm. The interface between Ta_2O_5 and Al_2O_3 layer was at 35 nm.

Figure 5b depicts the average hardness for triple-layered laminates. The hardness of the $Ta_2O_5/Al_2O_3/Ta_2O_5/Si$ laminate changed similarly along the depth to the wavelike shape of the indentation curve for the laminate $Al_2O_3/Ta_2O_5/Si$. Even though the system, in this case, is three-layered and consists mostly of the softer Ta_2O_5, the maximum was still quite high (9.6 GPa) at 15 nm and the minimum (9.3 GPa) at 25 nm. At the same time, Al_2O_3 layer begins at 24 nm and ends at 47 nm from the surface.

The second triple-layered laminate possessed a steady incline of hardness along the depth of the film (Figure 5b), which was similar to the behavior of the quadruple-layered laminate.

The summary of the hardness and moduli values of the films at three depths (10, 25, and 40 nm) are shown in Table 3. The surface was softer (8.2–9.5 GPa) for films with Ta_2O_5 top layer. The surface was the hardest at 11 GPa for the Al_2O_3/Si film and was lower by about 1 GPa for the $Al_2O_3/Ta_2O_5/Si$ film. At deeper displacements, the hardness values overlapped more significantly, yet the triple-layered laminate $Al_2O_3/Ta_2O_5/Al_2O_3/Si$ possessed slightly higher hardness compared to the rest of the laminates. Double- and quadruple-layered laminates had similar hardness at 25 and 40 nm displacements (10–11 GPa), while the triple-layer laminate with the largest amount of Ta_2O_5 was the softest among all the laminates.

Table 3. Average hardness and reduced modulus with standard deviation (SD) at various displacements.

Film	Hardness ± SD at 10 nm	Modulus ± SD at 10 nm	Hardness ± SD at 25 nm	Modulus ± SD at 25 nm	Hardness ± SD at 40 nm	Modulus ± SD at 40 nm
Al_2O_3/Si	11.2 ± 1.0	154 ± 9	12.3 ± 0.9	160 ± 6	13.1 ± 0.9	160 ± 6
Ta_2O_5/Si	8.2 ± 0.6	139 ± 5	8.9 ± 0.5	141 ± 4	9.7 ± 0.4	139 ± 3
$Ta_2O_5/Al_2O_3/Si$	8.2 ± 0.6	146 ± 6	10.3 ± 0.5	155 ± 3	10.9 ± 0.4	156 ± 2
$Al_2O_3/Ta_2O_5/Si$	10.4 ± 0.8	145 ± 6	9.9 ± 0.5	151 ± 4	10.4 ± 0.4	156 ± 4
$Al_2O_3/Ta_2O_5/Al_2O_3/Si$	9.9 ± 0.4	145 ± 4	10.5 ± 0.3	148 ± 3	11.1 ± 0.3	148 ± 3
$Ta_2O_5/Al_2O_3/Ta_2O_5/Si$	8.7 ± 0.9	147 ± 5	9.4 ± 0.6	151 ± 3	10.0 ± 0.5	153 ± 3
$Ta_2O_5/Al_2O_3/Ta_2O_5/Al_2O_3/Si$	9.5 ± 0.4	149 ± 3	10.0 ± 0.2	154 ± 2	10.7 ± 0.2	155 ± 2

According to the nanoindentation measurements, the elastic moduli of the laminates were similar and fell between the moduli of Ta_2O_5 and Al_2O_3 reference films (Table 3). The elastic modulus of the laminates did not indicate the presence of layers with different elasticities, while the difference between Al_2O_3/Si (~160 GPa) and Ta_2O_5/Si (~139 GPa) moduli was moderate, ~20 GPa.

Scanning probe microscopy images were taken of four indents on each sample, and an average descriptive image was chosen for each sample. Weak pile-up of 2 and 3 nm was observed on the substrate and the Al_2O_3 film, respectively, being only slightly higher than general surface roughness. However, Ta_2O_5 tended to pile up more significantly, up to 7 nm (not shown). The pile-up around the indents of the double-, triple- and quadruple-layered laminates was around 4 nm (Figure 6). Even for the $Ta_2O_5/Al_2O_3/Ta_2O_5/Si$ sample, the pile-up remained moderate, although it consisted mostly of Ta_2O_5, which tended to pile up most strongly. Intermediate Al_2O_3 layer decreased the pile-up behavior of the material.

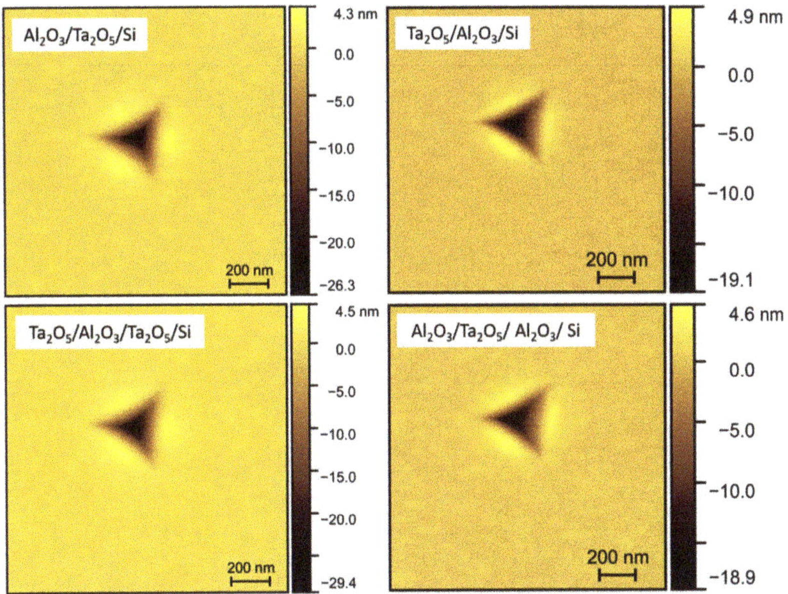

Figure 6. Scanning probe microscopy images of idents on double- and triple-layered laminates.

4. Discussion

Pelegri et al. [36] investigated theoretically the mechanical behavior of soft film-hard substrate and hard film-soft substrate systems during nanoindentation. The model materials possessed a difference in hardness within 27 GPa and resulted in hardness-displacement graphs wherein, due to the effect of the substrate, the hardness started to increase with displacement at 45% into the thickness of the soft film on hard substrate. At the same time, the hardness started to decrease only until displacement reached 30% of the thickness of the hard film on soft substrate. There was more pile-up occurring for the soft film-hard substrate system compared to the bulk soft material [36]. This approach could be used to explain the hardness-displacement curves of the double-layered laminates in the present study. The mechanical behaviour of the $Ta_2O_5/Al_2O_3/Si$ sample changed at around 20 nm displacement when the softer top Ta_2O_5 layer was penetrated about 60% (Figure 5a). Beginning from the displacement of about 33 nm, the hardness was slightly changed again, which is around 50% of the total laminate thickness. A displacement where the mechanical property varied could be considered where the underlying material started to influence the nanoindentation. Therefore, the hardness of 35 nm thick Ta_2O_5 began to be affected by the harder Al_2O_3 underlayer at 60% of the thickness of Ta_2O_5. The Ta_2O_5 layer decreased the overall hardness of the laminate, as the Ta_2O_5/Al_2O_3-Si system demonstrated a rather high difference in hardness along the indentation depth, and the substrate started to affect mechanical properties already at 50% of the thickness of the laminate. The depth where the substrate starts affecting the nanoindentation seemed to depend on the difference in the mechanical properties of the film and substrate materials.

The sample $Al_2O_3/Ta_2O_5/Si$ contained intrinsic transition from a harder material to softer material. Indentation until 30% of the top Al_2O_3 layer thickness, that is at 12 nm in depth, resulted in the maximum hardness value (Figure 5a). The displacement with the minimum hardness occurred at the depth of 34% of the total thickness of the laminate. The latter result indicates that, in the case of the Al_2O_3/Ta_2O_5-Si system, the substrate seems to affect measurement at smaller depths compared to the $Ta_2O_5/Al_2O_3/Si$ laminate, where the influence of the substrate became apparent at 50% of the thickness. Besides affecting the hardness near the surface, the sequence of the layers seemed to have some influence on the hardness near the substrate, too. In the latter case, however, the hardness of both double-layered laminates started to overlap at higher displacements (Figure 5a) (Table 3).

$Ta_2O_5/Al_2O_3/Ta_2O_5/Si$ and $Al_2O_3/Ta_2O_5/Si$ samples behaved similarly under the nanoindentation (Figure 5b). The difference between these laminates, as engineered, was the additional softer Ta_2O_5 top layer and the slightly lower layer thickness, so that the similar shape of the hardness-displacement curve could be attributed to the $Al_2O_3/Ta_2O_5/Si$ structure, which also constitutes the $Ta_2O_5/Al_2O_3/Ta_2O_5/Si$ film. The width of the peak on the curve was thinner for the triple-layered laminate compared to the double-layered one. This was probably due to the lower thickness of the layers. Complementarily, the additional Ta_2O_5 top layer or the indenter tip radius might affect the shape of the hardness-displacement curve. Lofaj et al. [37] found, both experimentally and theoretically, that the tip radius influences the resolution of the nanoindentation measurement in the case of thin films, i.e., to correctly determine the maximum hardness of a thin hard film on softer substrate the indenter tip radius must be sufficiently sharp. Hence, for nanolaminates with thin layers, the indenter tip radius might have an influence and must be taken into consideration. In this study, however, the indenter tip radius was undetermined, but the tip was calibrated (Figure 2) and gave correct results in the displacement range of 10 to 55 nm on the fused quartz glass.

In the case of the $Al_2O_3/Ta_2O_5/Al_2O_3/Si$ sample, the mechanically different layers remained undistinguishable on the hardness-displacement graph (Figure 5b). The Ta_2O_5 layer between two Al_2O_3 might have a smaller influence on the hardness as the deformation of the softer layer is impeded by the surrounding harder material, and as for the previous laminate, the harder layer can influence the deformation of the surrounding softer material. The quadruple-layered laminate acted similarly steady under nanoindentation conditions. However, in this case, the top Ta_2O_5 layer was not completely surrounded by the harder material, meaning the layer thickness also has an important influence on the mechanical properties of the laminates.

5. Conclusions

Amorphous Al_2O_3-Ta_2O_5 nanolaminates were atomic layers deposited on Si with an approximate thickness of 70 nm. The laminates possessed hardness in the range of 8.2–11.1 GPa. The hardness of the surface of the double-and triple-layered laminates was affected by the sequence of the layers, and quite possibly, the hardness at higher depths was also somewhat affected. The thickness of layers was also significant as the peak in the hardness-displacement curve appeared slimmer and vaguer as layer thickness decreased. Triple-layered $Al_2O_3/Ta_2O_5/Al_2O_3/Si$ laminate and quadruple-layered $Ta_2O_5/Al_2O_3/Ta_2O_5/Al_2O_3/Si$ laminate possessed a steady change of hardness with similar values along the depth, although consisted of various amounts of the oxides with a 3–4 GPa difference in hardness. The elastic modulus of amorphous Al_2O_3-Ta_2O_5 nanolaminates did not depend on the layer structure and fell between 145 and 155 GPa for all the laminates. The difference of hardness of the materials in the film-substrate systems seemed to affect the depth where the substrate started to influence nanoindentation. The beneficial effects of annealing on the mechanical behavior of the Al_2O_3-Ta_2O_5 double- and triple-layered nanolaminates will be investigated in a future study.

Author Contributions: Conceptualization, T.J. and H.-M.P.; methodology, P.R., T.J. and H.-M.P.; software, P.R., T.J. and H.-M.P.; validation, P.R., T.J. and H.-M.P.; resources, K.K. and A.T.; writing—original draft preparation, H.-M.P.; writing—review and editing, T.J., K.K. and A.T.; funding acquisition, K.K. and A.T. All authors have read and agreed to the published version of the manuscript.

Funding: The present study was partially funded by the European Regional Development Fund project "Emerging orders in quantum and nanomaterials" (Grant No. TK134) and Estonian Research Agency (Grant Nos. PRG4 and PRG753).

Institutional Review Board Statement: Not applicable.

Informed Consent Statement: Not applicable.

Data Availability Statement: The data that support the findings of this study are available from the corresponding author upon reasonable request.

Acknowledgments: Acknowledgements to Aivar Tarre, Aarne Kasikov, Hugo Mändar and Lauri Aarik for assisting with atomic layer deposition and film characterization measurements. This work was partially supported by the ERDF project "Center of nanomaterials technologies and research" (NAMUR+, Project No. 2014-2020.4.01.16-0123).

Conflicts of Interest: The authors declare no conflict of interest.

References

1. Jõgiaas, T.; Kull, M.; Seemen, H.; Ritslaid, P.; Kukli, K.; Tamm, A. Optical and mechanical properties of nanolaminates of zirconium and hafnium oxides grown by atomic layer deposition. *J. Vac. Sci. Technol. A* **2020**, *38*, 022406. [CrossRef]
2. Tamm, A.; Piirsoo, H.-M.; Jõgiaas, T.; Tarre, A.; Link, J.; Stern, R.; Kukli, K. Mechanical and magnetic properties of double layered nanostructures of tin and zirconium oxides grown by atomic layer deposition. *Nanomaterials* **2021**, *11*, 1633. [CrossRef] [PubMed]
3. Chu, K.; Shen, Y.G. Mechanical and tribological properties of nanostructured TiN/TiBN multilayer films. *Wear* **2008**, *265*, 516–524. [CrossRef]
4. Iantsunskyi, I.; Baitimitova, M.; Coy, E.; Yate, L.; Viter, R.; Ramanavičius, A.; Jurga, S.; Bechelany, M.; Erts, D. Influence of ZnO/graphene nanolaminate periodicity on their structural and mechanical properties. *J. Mater. Sci. Technol.* **2018**, *34*, 1487–1493. [CrossRef]
5. Abadias, G.; Uglov, V.V.; Saladukhin, I.A.; Zlotski, S.V.; Tolmachova, G.; Dub, S.N.; van Vuuren, A.J. Growth, structural and mechanical properties of magnetron-sputtered ZrN/SiN$_x$ nanolaminated coatings. *Surf. Coat. Technol.* **2016**, *308*, 158–167. [CrossRef]
6. Ghazaryan, L.; Handa, S.; Schmitt, P.; Beladiya, V.; Roddatis, V.; Tünnermann, A.; Szeghalmi, A. Tructural, optical, and mechanical properties of TiO$_2$ nanolaminates. *Nanotechnology* **2021**, *32*, 095709. [CrossRef]
7. Ylivaara, O.M.E.; Kilpi, L.; Liu, X. Aluminum oxide/titanium dioxide nanolaminates grown by atomic layer deposition: Growth and mechanical properties. *J. Vac. Sci. Technol. A* **2017**, *35*, 01B105. [CrossRef]
8. Coy, E.; Yate, L.; Kabacińska, Z.; Jancelewicz, M.; Jurga, S.; Iatsunskyi, I. Topographic reconstruction and mechanical analysis of atomic layer deposited Al$_2$O$_3$/TiO$_2$ nanolaminates by nanoindentation. *Mater. Des.* **2016**, *111*, 584–591. [CrossRef]
9. Kalam, K.; Rammula, R.; Ritslaid, P.; Käämbre, T.; Link, J.; Stern, R.; Vinuesa, G.; Dueñas, S.; Castán, H.; Tamm, A.; et al. Atomic layer deposited nanolaminates of zirconium oxide and manganese oxide from manganese (III) acetylacetonate and ozone. *Nanotechnology* **2021**, *32*, 335703. [CrossRef]
10. Kukli, K.; Kemell, M.; Castán, H.; Dueñas, S.; Link, J.; Stern, R.; Heikkilä, M.J.; Jõgiaas, T.; Kozlova, J.; Rähn, M.; et al. Magnetic properties and resistive switching in mixture films and nanolaminates consisting of iron and silicon oxides grown by atomic layer deposition. *J. Vac. Sci. Technol. A* **2020**, *38*, 042405. [CrossRef]
11. Kalam, K.; Seemen, H.; Mikkor, M.; Jõgiaas, T.; Ritslaid, P.; Tamm, A.; Kukli, K.; Kasikov, A.; Link, J.; Stern, R.; et al. Electrical and magnetic properties of atomic layer deposited cobalt oxide and zirconium oxide nanolaminates. *Thin Solid Films* **2019**, *669*, 294–300. [CrossRef]
12. Puurunen, R. Surface chemistry of atomic layer deposition: A case study for the trimethylaluminum/water process. *J. Appl. Phys.* **2005**, *97*, 121301. [CrossRef]
13. Gosset, L.G.; Damlencourt, J.-F.; Renault, O.; Rouchon, D.; Holliger, P.; Ermilieff, A. Interface and material characterization of thin Al$_2$O$_3$ layers deposited by ALD using TMA/H$_2$O. *J. Non-Cryst. Solids* **2002**, *303*, 12–23. [CrossRef]
14. Hanby, B.V.T.; Stuart, B.W.; Gimeno-Fabra, M.; Moffat, J.; Gerada, C.; Grant, D.M. Layered Al$_2$O$_3$-SiO$_2$ and Al$_2$O$_3$-Ta$_2$O$_5$ thin-film composites for high dielectric strength, deposited by pulsed direct current and radio frequency magnetron sputtering. *Appl. Surf. Sci.* **2019**, *492*, 328–336. [CrossRef]
15. Jenkins, M.A.; Austin, D.Z.; Holden, K.E.K.; Allman, D.; Conley, J.F. Laminate Al$_2$O$_3$/Ta$_2$O$_5$ Metal/Insulator/Insulator/Metal (MIIM) Devices for High-Voltage Applications. *IEEE Trans. Electron Devices* **2019**, *66*, 5260–5265. [CrossRef]
16. Song, W.; Wang, W.; Lee, H.K.; Li, M.; Zhuo, V.Y.-Q.; Chen, Z.; Chui, K.J.; Liu, J.-C.; Wang, I.-T.; Zhu, Y.; et al. Analog switching characteristics in TiW/Al$_2$O$_3$/Ta$_2$O$_5$/Ta RRAM devices. *Appl. Phys. Lett.* **2019**, *115*, 133501. [CrossRef]
17. Diaz, B.; Härkönen, E.; Swiatiwska, J.; Seyeux, A.; Maurice, V.; Ritala, M.; Marcus, P. Corrosion properties of steel protected by nanometre-thick oxide coatings. *Corros. Sci.* **2014**, *82*, 208–217. [CrossRef]
18. Kukli, K.; Kemell, M.; Vehkamäki, M.; Heikkilä, M.J.; Mizohata, K.; Kalam, K.; Ritala, M.; Leskelä, M.; Kundrata, I.; Fröhlich, K. Atomic layer deposition and properties of mixed Ta$_2$O$_5$ and ZrO$_2$ films. *AIP Adv.* **2017**, *7*, 025001. [CrossRef]
19. Kukli, K.; Ritala, M.; Leskelä, M.; Sajavaara, T.; Keinonen, J.; Gilmer, D.; Bagchi, S.; Prabhu, L. Atomic layer deposition of Al$_2$O$_3$, ZrO$_2$, Ta$_2$O$_5$, and Nb$_2$O$_5$ based nanolayered dielectrics. *J. Non-Cryst. Solids* **2002**, *303*, 35–39. [CrossRef]
20. Kukli, K.; Ritala, M.; Leskelä, M. Development of dielectric properties of niobium oxide, tantalum oxide, and aluminum oxide based nanolayered materials. *J. Electrochem. Soc.* **2001**, *148*, 35–41. [CrossRef]
21. Kukli, K.; Ritala, M.; Leskelä, M. Properties of atomic layer deposited (Ta$_{1-x}$Nb$_x$)$_2$O$_5$ solid solution films and Ta$_2$O$_5$-Nb$_2$O$_5$ nanolaminates. *J. Appl. Phys.* **1999**, *86*, 5656–5662. [CrossRef]

22. Kukli, K.; Ritala, M.; Leskelä, M. Atomic layer epitaxy growth of tantalum oxide thin films from Ta(OC$_2$H$_5$)$_5$ and H$_2$O. *J. Electrochem. Soc.* **1995**, *142*, 1670–1675. [CrossRef]
23. Tapily, K.; Jakes, J.E.; Stone, D.S.; Shrestha, P.; Gu, D.; Baumgart, H.; Elmustafa, A.A. Nanoindentation investigation of HfO$_2$ and Al$_2$O$_3$ films grown by atomic layer deposition. *J. Electrochem. Soc.* **2008**, *155*, 545–551. [CrossRef]
24. Jõgiaas, T.; Zables, R.; Tarre, A.; Tamm, A. Hardness and modulus of elasticity of atomic layer deposited Al$_2$O$_3$-ZrO$_2$ nanolaminates and mixtures. *Mater. Chem. Phys.* **2020**, *240*, 1222270. [CrossRef]
25. Piirsoo, H.-M.; Jõgiaas, T.; Mändar, H.; Ritslaid, P.; Kukli, K.; Tamm, A. Microstructure and mechanical properties of atomic layer deposited alumina doped zirconia. *AIP Adv.* **2021**, *11*, 055316. [CrossRef]
26. Jõgiaas, T.; Zables, R.; Tamm, A.; Merisalu, M.; Hussainova, I.; Heikkilä, M.; Mändar, H.; Kukli, K.; Ritala, M.; Leskelä, M. Mechanical properties of aluminum, zirconium, hafnium and tantalum oxides and their nanolaminates grown by atomic layer deposition. *Surf. Coat. Technol.* **2015**, *282*, 36–42. [CrossRef]
27. Borylo, P.; Lukaszkowicv, K.; Szindler, M.; Kubacki, J.; Balin, K.; Basiaga, M.; Szewczenko, J. Structure and properties of Al$_2$O$_3$ thin films deposited by ALD process. *Vacuum* **2016**, *131*, 319–326. [CrossRef]
28. Kukli, K.; Salmi, E.; Jõgiaas, T.; Zabels, R.; Schuisky, M.; Westlinder, J.; Mizohata, K.; Ritala, M.; Leskelä, M. Atomic layer deposition of aluminum oxide on modified steel substrates. *Surf. Coat. Technol.* **2016**, *304*, 1–8. [CrossRef]
29. Băilă, D.-I.; Vitelaru, C.; Truscă, R.; Constantin, L.R.; Păcurar, A.; Parau, C.A.; Păcurar, R. Thin films deposition of Ta$_2$O$_5$ and ZnO by e-gun technology on Co-Cr alloy manufactured by direct metal laser sintering. *Materials* **2021**, *14*, 3666. [CrossRef]
30. Rahmati, B.; Sarhan, A.A.D.; Zalnezhad, E.; Kamiab, Z.; Dabbagh, A.; Choudhury, D.; Abas, W.A.B.W. Development of tantalum oxide (Ta-O) thin film coating on biomedical Ti-6Al-4V alloy to enhance mechanical properties and biocompatibility. *Ceram. Int.* **2016**, *42*, 466–480. [CrossRef]
31. Abernathy, M.R.; Hough, J.; Martin, I.W.; Rowan, S.; Oyen, M.; Linn, C.; Faller, J.E. Investigation of the Young's modulus and thermal expansion of amorphous titania-doped tantala films. *Appl. Opt.* **2014**, *53*, 3196–3202. [CrossRef] [PubMed]
32. Arroval, T.; Aarik, L.; Rammula, R.; Kruusla, V.; Aarik, J. Effect of substrate-enhanced and inhibited growth on atomic layer deposition and properties of aluminum–titanium oxide films. *Thin Solid Films* **2016**, *600*, 119–125. [CrossRef]
33. Mändar, H.; Felsche, J.; Mikli, V.; Vajakas, T. AXES1.9: New tools for estimation of crystallite size and shape by Williamson–Hall analysis. *J. Appl. Crystallogr.* **1999**, *32*, 345–350. [CrossRef]
34. Oliver, W.C.; Pharr, G.M. Measurement of hardness and elastic modulus by instrumented indentation: Advances in understanding and refinements to methodology. *J. Mater. Res.* **2004**, *19*, 3–20. [CrossRef]
35. Ma, Z.S.; Zhou, Y.C.; Long, S.G.; Lu, C. On the intrinsic hardness of a metallic film/substrate system: Indentation size and substrate effects. *Int. J. Plast.* **2012**, *34*, 1–11. [CrossRef]
36. Pelegri, A.A.; Juang, X. Nanoindentation on soft film/hard substrate and hard film/soft substrate material systems with finite element analysis. *Compos. Sci. Technol.* **2008**, *68*, 147–155. [CrossRef]
37. Lofaj, F.; Németh, D. The effects of tip sharpness and coating thickness on nanoindentation measurements in hard coatings on softer substrates by FEM. *Thin Solid Films* **2017**, *644*, 173–181. [CrossRef]

Article

Numerical Investigation for the Resin Filling Behavior during Ultraviolet Nanoimprint Lithography of Subwavelength Moth-Eye Nanostructure

Yuanchi Cui [1,2], Xuewen Wang [1,2], Chengpeng Zhang [1,3,4,*], Jilai Wang [1] and Zhenyu Shi [1]

[1] Key Laboratory of High Efficiency and Clean Mechanical Manufacture of Ministry of Education, School of Mechanical Engineering, National Demonstration Center for Experimental Mechanical Engineering Education, Shandong University, Jinan 250061, China; cuiyuanchi@mail.sdu.edu (Y.C.); xwwang1990@mail.sdu.edu (X.W.); jlwang@sdu.edu.cn (J.W.); shizhenyu@sdu.edu.cn (Z.S.)

[2] Nanjing Rolling Intelligent Manufacturing Research Institute, Nanjing Mumuxili Technology Co., Ltd., Nanjing 211100, China

[3] State Key Laboratory of Mechanical System and Vibration, Shanghai Jiao Tong University, Shanghai 200240, China

[4] Suzhou Institute of Shandong University, Shandong University, Suzhou 215123, China

* Correspondence: zhangchengpeng@sdu.edu.cn

Abstract: Accurate analysis of the resin filling process into the mold cavity is necessary for the high-precision fabrication of moth-eye nanostructure using the ultraviolet nanoimprint lithography (UV-NIL) technique. In this research, a computational fluid dynamics (CFD) simulation model was proposed to reveal resin filling behavior, in which the effect of boundary slip was considered. By comparison with the experimental results, a good consistency was found, indicating that the simulation model could be used to analyze the resin filling behavior. Based on the proposed model, the effects of process parameters on resin filling behavior were analyzed, including resin viscosity, inlet velocity and resin thickness. It was found that the inlet velocity showed a more significant effect on filling height than the resin viscosity and thickness. Besides, the effects of boundary conditions on resin filling behavior were investigated, and it was found the boundary slip had a significant influence on resin filling behavior, and excellent filling results were obtained with a larger slip velocity on the mold side. This research could provide guidance for a more comprehensive understanding of the resin filling behavior during UV-NIL of subwavelength moth-eye nanostructure.

Keywords: CFD simulation; ultraviolet nanoimprint lithography; filling behavior; moth-eye nanostructure

Citation: Cui, Y.; Wang, X.; Zhang, C.; Wang, J.; Shi, Z. Numerical Investigation for the Resin Filling Behavior during Ultraviolet Nanoimprint Lithography of Subwavelength Moth-Eye Nanostructure. *Coatings* **2021**, *11*, 799. https://doi.org/10.3390/coatings11070799

Received: 30 May 2021
Accepted: 22 June 2021
Published: 1 July 2021

Publisher's Note: MDPI stays neutral with regard to jurisdictional claims in published maps and institutional affiliations.

1. Introduction

The moth-eye nanostructure has been widely used to reduce the surface reflection on optoelectronic products, such as liquid crystal display (LCD) [1,2], solar cell [3,4], image sensor [5], etc. Inspired by nature, the moth-eye nanostructure is a nipple array, typically of nano-scale height and diameter [6]. Its antireflection performance is closely related to structural accuracy, and therefore high-precision fabrication of moth-eye nanostructure is critical [7]. Among various fabrication technologies, the ultraviolet nanoimprint lithography (UV-NIL) provides an effective solution for the high-precision fabrication of moth-eye nanostructure [8,9], which mainly includes liquid resin filling into the mold cavity and UV-curing process. The filling process is core during UV-NIL of moth-eye nanostructure, which determines the final forming accuracy, so an accurate analysis of the resin filling behavior is necessary [10,11]. However, the moth-eye nanostructure is generally tens to hundreds of nanometers [12,13], and it is quite difficult to analyze the resin filling behavior through the experimental method. Therefore, numerical simulation is commonly adopted to investigate the resin filling behavior in micro/nano-scale [14,15].

Various researches have been carried out to investigate the resin filling behavior in the UV-NIL process. Kim et al. [16] studied the effects of contact angle and aspect ratio on the resin filling behavior, and it was found that complete filling occurred when the contact angle on the vertical wall was as low as that on the substrate. Lai et al. [17,18] investigated the effects of process parameters on bubble defects during roll-to-roll ultraviolet nanoimprint lithography (R2R UV-NIL) of the micro-pyramid array. Tian et al. [19] proposed a simulation model to predict the electrically induced pattern formation, which considered the effects of electrostatic force-assisted and electrocapillary force-driven. Nagaoka et al. [20] studied the resin filling characteristics under a condensable gas ambient, and it was found the gas ambient had a significant impact on the resin filling behavior. Shibata et al. [21] established a simulation system to investigate the UV-NIL process, which was composed of the resin filling module, optical-intensity distribution module, mechanical properties module and shrinkage module. Ye et al. [22] analyzed the generation mechanism of bubble defects in the R2R UV-NIL process.

Although various researches have been reported to analyze resin filling behavior, the moth-eye nanostructure is generally tens to hundreds of nanometers and boundary slip has a significant effect on the filling behavior at this scale [23,24]. Few simulation studies have been reported considering the effect of boundary slip. The computational fluid dynamics (CFD) simulation is an effective method to predict the behavior of liquids and gases in an intelligent way, which is widely used in aerospace, automotive, energy and other fields. Therefore, a CFD simulation model was proposed in this paper, which considered the effect of boundary slip. Based on the proposed model, the effects of process parameters and boundary conditions on resin filling behavior were investigated. Besides, the simulation results were also compared with experimental results, and good agreement was found. This research could provide guidance for the accurate analysis of resin filling behavior during UV-NIL of moth-eye nanostructure.

2. Numerical Modeling

2.1. Physical Model

Figure 1a shows the schematic diagram of resin filling process during UV-NIL of moth-eye nanostructure, where v is flow velocity of the resin, d the diameter of moth-eye nanostructure, h the height of moth-eye nanostructure and L the resin thickness. The imprinting mold for moth-eye nanostructure is 90 nm in diameter and 180 nm in depth, as shown in Figure 1b. A simulation model was built to investigate the resin filling behavior during UV-NIL of moth-eye nanostructure using the commercial program, ANSYS-Fluent, and the following assumptions were made to simplify the simulation process.

(1) As shown in Figure 1b, the moth-eye nanostructure was rotationally symmetric, and therefore the simulation model was simplified to a two-dimensional model.
(2) The resin filling process was assumed as an incompressible steady laminar flow during UV-NIL of moth-eye nanostructure [16,25].
(3) The resin was assumed as an incompressible Newton fluid [20,26].
(4) It was assumed that the resin flowed into the gap between the mold and substrate from one side at a certain speed and out from the other side [25].

2.2. Material Parameters

The material parameters could obviously affect filling behavior, so an accurate setting of material parameters in the simulation model was necessary. Two kinds of materials were included in the simulation model, that was air and resin. As one common material, the material parameters for air were chosen from a database in ANSYS-Fluent software (R15.0). The resin for fabrication of moth-eye nanostructure was a radical curing system developed by Jiangsu Kangde Xin Composite Material Co., Ltd. (Suzhou, China) [7,27] and its material parameters were obtained through experimental characterization, including viscosity, surface tension and wettability.

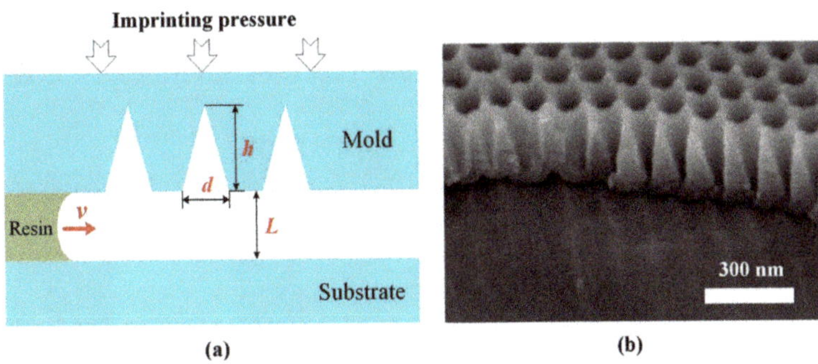

Figure 1. The UV-NIL of moth-eye nanostructure (**a**) schematic diagram (**b**) imprinting mold.

Figure 2 presented the viscosity property at different temperatures, which was obtained using the rotary viscometer (DV2T, Brookfield, Middleboro, MA, USA). It was found that the resin viscosity decreased as the temperature increased. When the temperature was raised from 25 °C to 65 °C, the resin viscosity was reduced from 322.3 mPa·s to 45.6 mPa·s. This phenomenon could be attributed to the thermal motion energy of polymer molecules that was larger with a higher temperature, which made the liquid internal pores increase in both number and volume, so the resin fluidity was better [18].

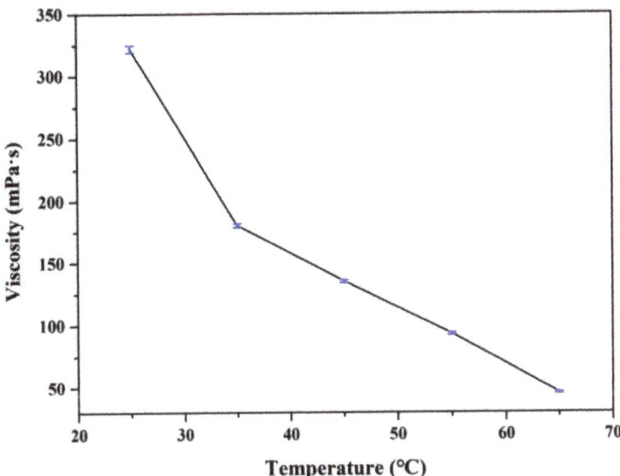

Figure 2. The resin viscosity at different temperatures.

Figure 3 displayed the surface tension at different temperatures, which was obtained with a surface tension tester (K100, Kruss, Hamburg, Germany) and Wilhelmy method. The temperature was set as 25 °C, 50 °C, 75 °C, respectively. As shown in Figure 3, the surface tension varied from 29.0 mN/m to 25.0 mN/m as the temperature was raised from 25 °C to 75 °C. The reason for this phenomenon was that the thermal motion became more intense inside the liquid, and the distance between molecules increased when the temperature rose, which caused the gravitational pull from the surface molecules to decrease accordingly. Moreover, the density of the gas phase and the gas stress relative to the surface molecules increased when the temperature rose. The two effects worked together, which caused surface tension to decrease at high temperatures [17].

Figure 4 gave the contact angle measurements, which were obtained using the contact angle measuring instrument (DSA30, Kruss, Hamburg, Germany). The volume of the droplet for contact angle measurements was 3 μL. In order to ensure accuracy, at least five different positions were chosen to measure the contact angle on each sample surface, and the average of all measurements was taken as the contact angle of the sample surface.

Based on the measurements, the contact angle between resin with mold and polyethylene terephthalate (PET) substrate was $21.5 \pm 1.8°$ and $28.3 \pm 2.1°$, respectively.

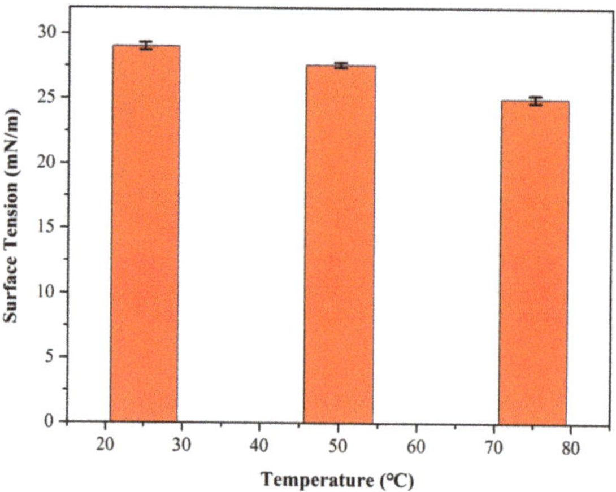

Figure 3. The surface tension of the resin at different temperatures.

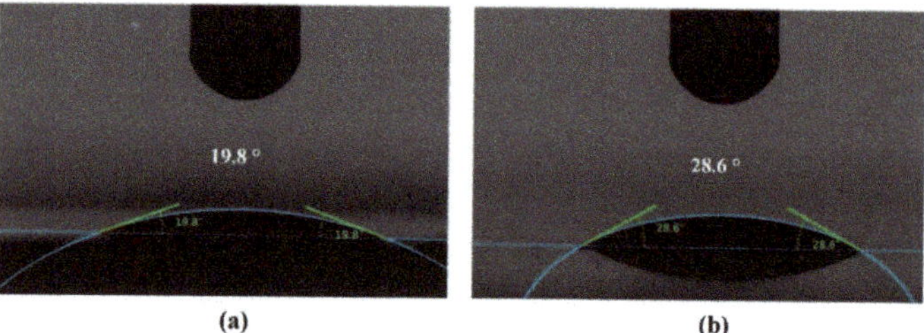

(a)   (b)

Figure 4. The contact angle between the resin and (**a**) mold (**b**) substrate.

Based on the experimental characterization above, the parameters in the simulation model were set as Table 1.

Table 1. Physical parameters in the filling simulation model.

Physical Constant	Values
Viscosity of the resin	$0.05{\sim}0.35$ Pa·s
Density of the resin	1200 kg/m^3
Viscosity of the air	1.79×10^{-5} Pa·s
Density of the air	1.225 kg/m^3
Surface tension of the resin	$25{\sim}29$ mN/m
Contact angle between the resin and substrate	$28.3 \times 2.1°$
Contact angle between the resin and mold	$21.5 \times 1.8°$

2.3. Boundary Conditions

Figure 5 presented the boundary conditions in the simulation model. As shown in Figure 1a, the resin flowed into the gap between the mold and substrate from one side at a certain speed and out from the other side. Therefore, the left side was considered to be the resin inlet, and the boundary condition was set as velocity inlet. The right side was considered to be the resin outlet, and the boundary condition was set as outflow. The top side was the impermeable mold surface, and the boundary condition was set as a wall. The bottom side was impermeable PET substrate, and the boundary condition was set as a wall.

According to whether the slip occurred, the wall boundary condition could be further set as no-slip wall and slip wall.

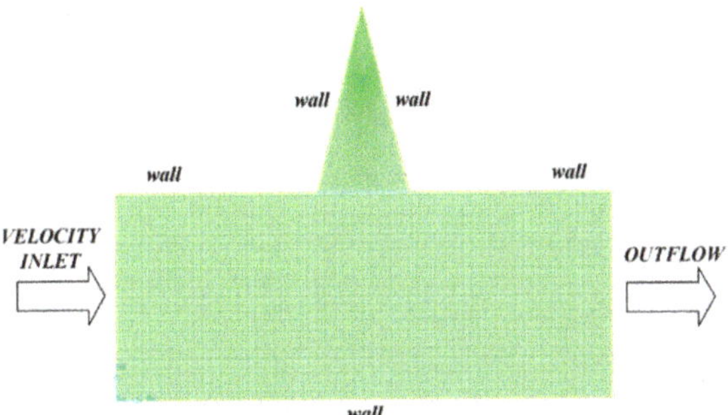

Figure 5. Boundary conditions in the simulation model.

2.4. Mathematic Equations

The Navier–Stokes equation was used to describe the resin flow behavior. Mass conservation was given by

$$\frac{\partial u}{\partial x} + \frac{\partial v}{\partial y} = 0 \tag{1}$$

where u was the velocity in x direction and v was the velocity in y direction.

Momentum conservation was given by

$$\rho\left(u\frac{\partial u}{\partial x} + v\frac{\partial u}{\partial y}\right) = -\frac{\partial p}{\partial x} + \mu\left(\frac{\partial^2 u}{\partial x^2} + \frac{\partial^2 u}{\partial y^2}\right) \tag{2}$$

$$\rho\left(u\frac{\partial v}{\partial x} + v\frac{\partial v}{\partial y}\right) = -\frac{\partial p}{\partial y} + \mu\left(\frac{\partial^2 v}{\partial x^2} + \frac{\partial^2 v}{\partial y^2}\right) \tag{3}$$

where p was the pressure, ρ the fluid density and μ the fluid viscosity.

In the simulation model, the air was set as the primary phase, and the resin was set as the secondary phase. The volume of fluid (VOF) method was employed to track the interface between resin and air [28,29]. The VOF function $f_{i,j}$ was defined in the entire flow field, which was the ratio of fluid volume to mesh volume in each mesh. A mesh that did not contain this fluid was called an empty mesh, a mesh that was filled with such a fluid was called a full mesh, and a mesh that contained an interface was called a half mesh. Assuming the calculation area was Ω and the fluid area was Ω_l, the following function was defined:

$$f(x,t) = \begin{cases} 1 & x \in \Omega_l \\ 0 & x \notin \Omega_l \end{cases} \tag{4}$$

The equation $\frac{\partial f(x,t)}{\partial t} + u\frac{\partial f(x,t)}{\partial x} + v\frac{\partial f(x,t)}{\partial y} = 0$ was satisfied in the flow field, and $V = (u,v)$ was the velocity field. The VOF function $f_{i,j}$ was defined as an integral of $f(x,t)$ on each mesh $I_{i,j}$.

$$f_{i,j} = \frac{1}{\Delta V_{i,j}}\int f(x,t)dV \tag{5}$$

The mesh with $f_{i,j} = 1$ was full of fluid, the mesh with $f_{i,j} = 0$ was called the empty mesh and the mesh with $0 < f_{i,j} < 1$ was called the boundary mesh. According to the $f_{i,j}$ value, the interface could be constructed.

2.5. Evaluation Method

The filling height was adopted to describe the simulation results quantitatively, as shown in Figure 6. After the filling process was completed, the simulation results were post-processed, and then the resin volume fraction in the mold cavity was derived. According to the cloud image of volume fraction, the interface between resin and air could be extracted. In this research, the height value at the mold center line was defined as filling height.

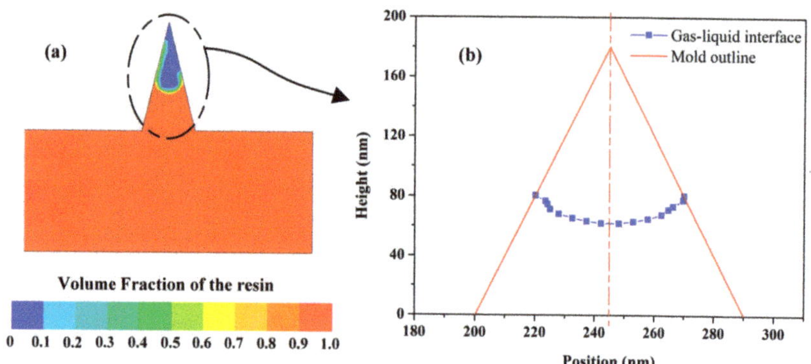

Figure 6. The filling height of moth-eye nanostructure from the simulation results. (**a**) the resin volume fraction in the mold cavity, (**b**) the interface between resin and air.

3. Results and Discussion

3.1. Effect of Process Parameters

Figure 7 showed the effect of resin viscosity on the filling height of the moth-eye nanostructure. In order to ensure one single variable, the inlet velocity and resin thickness were set as 1×10^{-5} m/s and 200 nm. As observed, the filling height decreased as the resin viscosity increased both with no-slip and slip boundaries. The reason for this phenomenon was that as the resin viscosity increased, the viscous resistance increased and less resin flowed into the mold cavity under the same conditions, thereby obtaining a lower filling height. Figure 8 displayed the moth-eye nanostructures fabricated by the UV-NIL process, and geometric characterization was conducted using atomic force microscope (AFM) equipment (Dimension FastScan, Bruker, Karlsruhe, Germany). As presented in Figure 8c, the forming height was acquired through bearing analysis of AFM images. It could be found that the simulation results with slip boundary were more compatible with the experimental results. The experimental results were slightly lower than the simulation heights, which might be caused by the curing shrink of the resin [30,31]. Based on the contrast between simulation and experimental results, the proposed simulation model considering the effect of boundary slip showed high accuracy, which could be used for the analysis of resin filling behavior during the UV-NIL of moth-eye nanostructure.

Figure 9 showed the effect of inlet velocity on the filling height of the moth-eye nanostructure. According to the simulation results, the filling height declined as the inlet velocity increased. It could be attributed to the mold cavity that was closed more quickly as the inlet velocity increased so that the resin volume flowing into the mold cavity decreased, eventually causing the height to decrease. Further analysis showed that a larger filling height could be obtained with slip boundary, which was because the resin near the mold side flowed faster than the substrate side, and more resin could be filled into the mold cavity before the cavity was closed. However, when the inlet velocity was larger than 7×10^{-5} m/s, the filling height was almost the same with both slip and no-slip boundaries. This was because the filling height was affected by the boundary and initial condition simultaneously. When the initial inlet velocity became larger, the influence of boundary slip was weakened, and the final height was mainly affected by the inlet velocity [32,33].

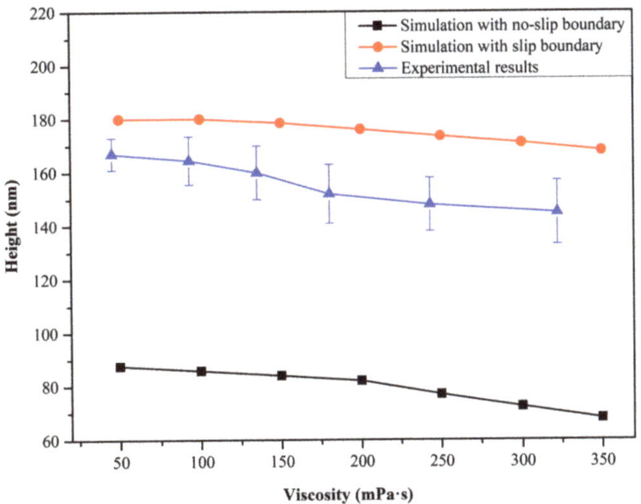

Figure 7. Effect of resin viscosity on the filling height of moth-eye nanostructure.

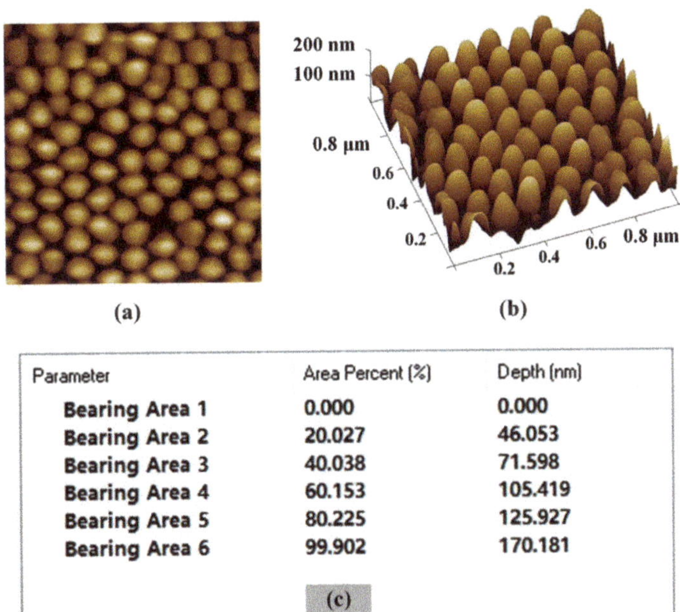

Parameter	Area Percent (%)	Depth (nm)
Bearing Area 1	0.000	0.000
Bearing Area 2	20.027	46.053
Bearing Area 3	40.038	71.598
Bearing Area 4	60.153	105.419
Bearing Area 5	80.225	125.927
Bearing Area 6	99.902	170.181

(c)

Figure 8. AFM characterization of the fabricated moth-eye nanostructure (**a**) two-dimensional topography (**b**) three-dimensional morphology (**c**) bearing analysis.

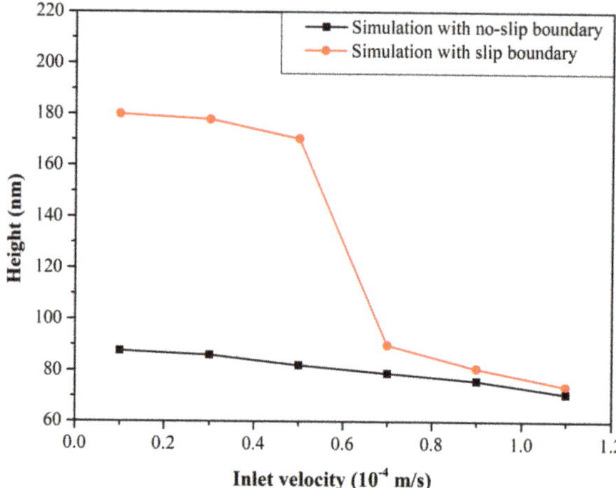

Figure 9. Effect of inlet velocity on the filling height of moth-eye nanostructure.

Figure 10 showed the effect of resin thickness on the filling height of moth-eye nanostructure. As observed, the filling height decreased as the resin thickness increased. It was because the resin thickness was adjusted by the imprinting pressure and smaller resin thickness meant a larger pressure. When the resin thickness became larger, the pressure decreased, and less resin was squeezed into the mold cavity, thereby causing a lower filling height. Besides, it was further found that a larger filling height could be obtained with slip boundary, which was because the slip on the mold caused the resin near the mold side to flow faster than the substrate side, and more resin could be filled into the mold cavity to obtain a larger filling height before the cavity was closed.

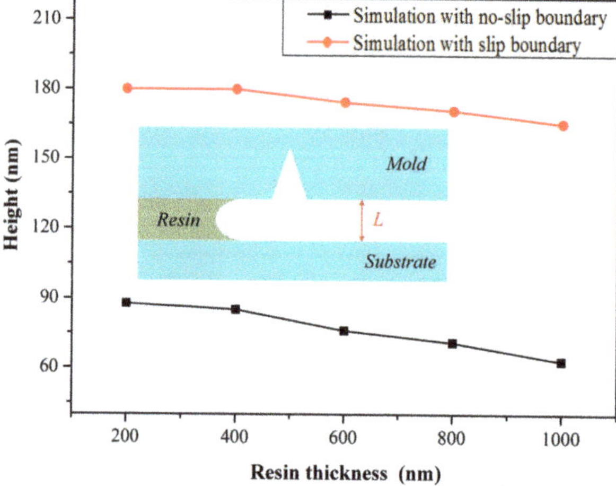

Figure 10. Effect of resin thickness on the filling height of the moth-eye nanostructure.

3.2. Effect of Slip Position

The boundary slip might occur on different positions, substrate side or mold side, and it was difficult to investigate the influence of slip position on filling behavior using the experimental method. Therefore, the effect of slip position on filling behavior was investigated based on the simulation model, as shown in Figure 11. In order to ensure one single variable, the slip speed was set to the same value, which was 2.5×10^{-5} m/s. According to the simulation results, the slip position showed a significant influence on the resin filling behavior, and the forming height was 87.6 nm, 73.2 nm, 54.3 nm and 180 nm with no-slip on both the mold and the substrate, slip on both the mold and the substrate, slip on the substrate only and slip on the mold only, respectively. Based on the results, it was concluded that the solid–liquid interface property had an important influence on the resin filling process, which affected the slip position.

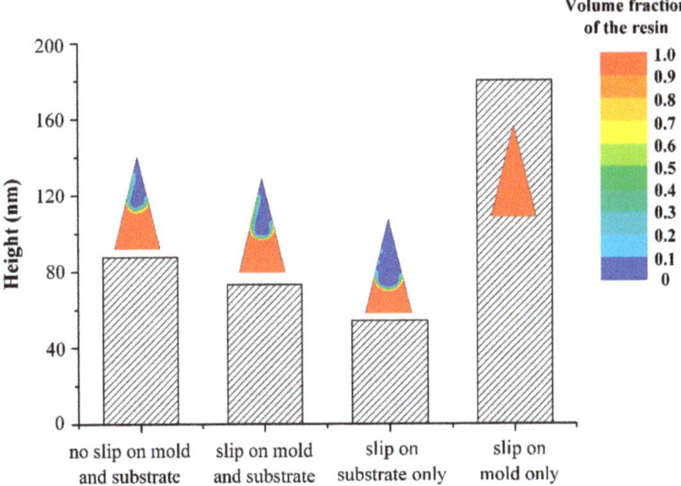

Figure 11. Effect of slip position on the filling height of moth-eye nanostructure.

The resin filling evolution process with different slip positions was further investigated, as shown in Figure 12. It could be found that the resin near the substrate flowed faster when boundary slip occurred on the substrate side only, so the cavity was closed more quickly, thereby lowering the filling height. Besides, it was observed from Figure 12d that the resin near the mold side flowed faster when the boundary slip occurred on the mold side only, and more resin could flow into the mold cavity before the cavity was closed, thereby obtaining a larger filling height.

3.3. Effect of Slip Velocity

Figure 13 showed the effect of slip velocity on the filling height of the moth-eye nanostructure. In order to ensure one single variable, the slip positions were all on the mold side only, the inlet velocity was 1.0×10^{-5} m/s, the resin viscosity was 0.1 Pa·s and the resin thickness was 200 nm, respectively. It could be observed that the filling height increased as the slip velocity increased. However, the influence of slip velocity was not significant when the slip velocity was less than 1.0×10^{-5} m/s, while the filling height was significantly affected when the slip velocity was larger than 1.0×10^{-5} m/s. The reason for this phenomenon was that the filling height was affected by the slip velocity and resin inlet velocity simultaneously, and the filling height was mainly affected by the resin inlet velocity when the slip velocity was small while it was mainly affected by the slip velocity when the slip velocity was larger than a certain threshold [32,33].

Figure 14 showed the filling evolution process with different slip velocity, which was 0 m/s, 1.0×10^{-5} m/s and 2.0×10^{-5} m/s, respectively. It was observed that the flow velocity near the mold side increased as the slip velocity increased so that more resin could flow into the mold cavity under the same conditions, thereby obtaining a larger filling height.

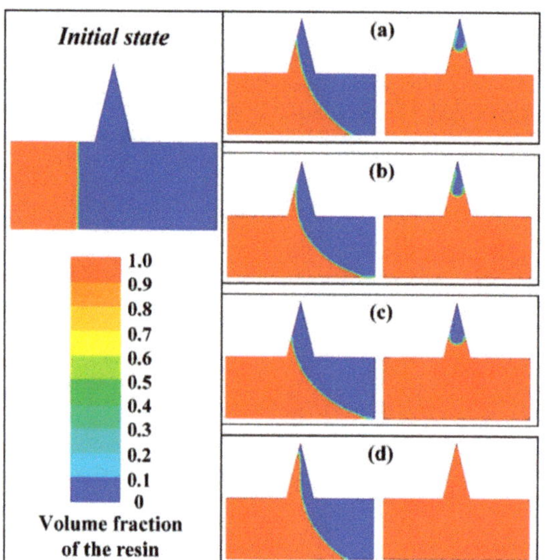

Figure 12. The filling evolution process with different slip positions (**a**) no-slip on both the mold and the substrate (**b**) slip on both the mold and the substrate (**c**) slip on the substrate only (**d**) slip on the mold only.

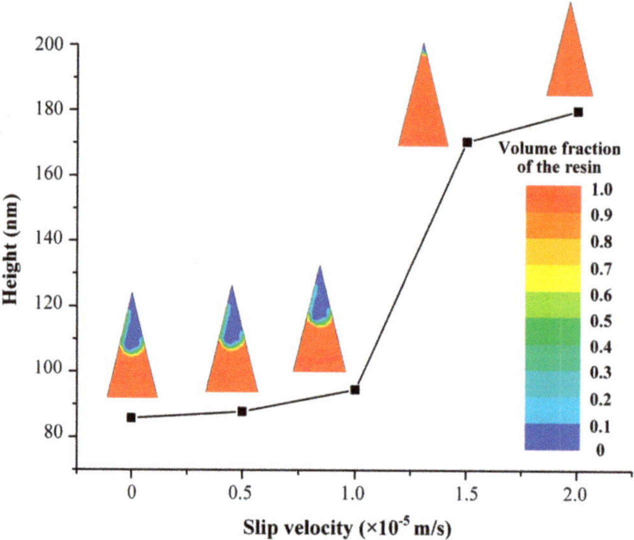

Figure 13. The effect of slip velocity on the filling height of the moth-eye nanostructure.

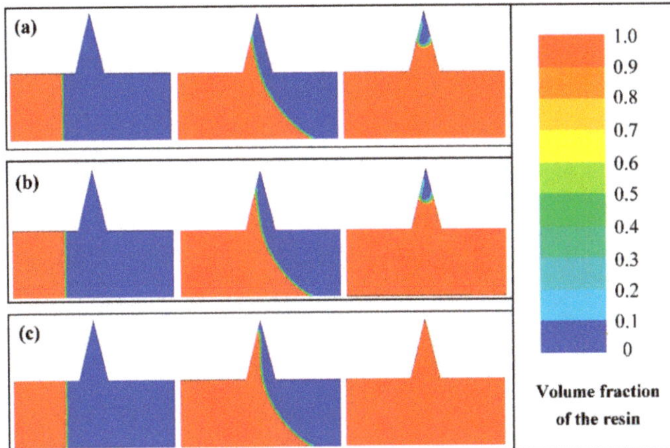

Figure 14. The resin filling evolution process with different slip velocity (**a**) 0 m/s (**b**) 1.0×10^{-5} m/s (**c**) 2.0×10^{-5} m/s.

4. Conclusions

A CFD simulation model was established to investigate the resin filling behavior during the UV-NIL process of the moth-eye nanostructure, where the effect of boundary slip was considered. By comparison with the experimental results, a good consistency was found, indicating that the simulation model could be used to analyze the resin filling behavior in the UV-NIL process of the moth-eye nanostructure.

Based on the proposed model, the effects of process parameters on filling behavior were analyzed, including resin viscosity, inlet velocity and resin thickness. It was found that the inlet velocity showed a more significant effect on filling height than resin viscosity and thickness. Therefore, choosing a reasonable speed was quite important during the UV-NIL of the moth-eye nanostructure, taking into account the forming accuracy and efficiency comprehensively, which could provide guidance for the UV-NIL experiments of the moth-eye nanostructure.

Besides, the effects of boundary conditions on the mold filling behavior of moth-eye nanostructure were also investigated. It was found that the slip position showed a significant influence on the resin filling behavior and the forming height was 87.6 nm, 73.2 nm, 54.3 nm and 180 nm with no-slip on both the mold and the substrate, slip on both the mold and the substrate, slip on the substrate only, and slip on the mold only, respectively. Moreover, the filling height was affected by the slip velocity and resin inlet velocity simultaneously, and the filling height was mainly affected by the inlet velocity when the slip velocity was small while it was mainly affected by the slip velocity when the slip velocity was larger than a certain threshold. The results could provide guidance for the regulation of solid–liquid interface properties.

Author Contributions: Drafting the work, Y.C., X.W.; The acquisition and analysis of data for the work, C.Z., J.W. and Z.S. All authors have read and agreed to the published version of the manuscript.

Funding: This research received no external funding.

Institutional Review Board Statement: Not applicable.

Informed Consent Statement: Not applicable.

Data Availability Statement: Data is contained within the article.

Acknowledgments: The authors would like to thank the Natural Science Foundation of Jiangsu Province (BK20190202), the China Postdoctoral Science Foundation (2021M691926), the Research Project of State Key Laboratory of Mechanical System and Vibration (MSV202102), the Shandong Provincial Natural Science Foundation (ZR2019BEE062) and Young Scholars Program of Shandong University.

Conflicts of Interest: The authors declare no competing financial interests.

References

1. Chen, H.; Lan, Y.F.; Tsai, C.Y.; Wu, S.T. Low-voltage blue-phase liquid crystal display with diamond-shape electrodes. *Liq. Cryst.* **2016**, *44*, 1124–1130. [CrossRef]
2. Liu, Y.; Lai, J.; Li, X.; Xiang, Y.; Li, J.; Zhou, J. A quantum dot array for enhanced tricolor liquid-crystal display. *IEEE Photonics J.* **2017**, *9*, 1–7. [CrossRef]
3. Yoshikawa, K.; Kawasaki, H.; Yoshida, W.; Irie, T.; Konishi, K.; Nakano, K.; Uto, T.; Adachi, D.; Kanematsu, M.; Uzu, H.; et al. Silicon heterojunction solar cell with interdigitated back contacts for a photoconversion efficiency over 26%. *Nat. Energy* **2017**, *2*, 17032. [CrossRef]
4. Zheng, Z.; Awartani, O.M.; Gautam, B.; Liu, D.; Qin, Y.; Li, W.; Bataller, A.; Gundogdu, K.; Ade, H.; Hou, J. Efficient charge transfer and fine-tuned energy level alignment in a THF-processed fullerene-free organic solar cell with 11.3% efficiency. *Adv. Mater.* **2017**, *29*, 1604241. [CrossRef] [PubMed]
5. Goossens, S.; Navickaite, G.; Monasterio, C.; Gupta, S.; Piqueras, J.J.; Pérez, R.; Burwell, G.; Nikitskiy, I.; Lasanta, T.; Galán, T.; et al. Broadband image sensor array based on graphene–CMOS integration. *Nat. Photonics* **2017**, *11*, 366–371. [CrossRef]
6. Vukusic, P.; Sambles, J.R. Photonic structures in biology. *Nat. Cell Biol.* **2003**, *424*, 852–855. [CrossRef] [PubMed]
7. Zhang, C.; Yi, P.; Peng, L.; Ni, J. Optimization and continuous fabrication of moth-eye nanostructure array on flexible polyethylene terephthalate substrate towards broadband antireflection. *Appl. Opt.* **2017**, *56*, 2901–2907. [CrossRef] [PubMed]
8. Zhang, C.; Yi, P.; Peng, L.; Lai, X.; Ni, J. Fabrication of moth-eye nanostructure arrays using roll-to-roll uv-nanoimprint lithography with an anodic aluminum oxide mold. *IEEE Trans. Nanotechnol.* **2015**, *14*, 1127–1137. [CrossRef]
9. Brousseau, E.B.; Dimov, S.S.; Pham, D. Some recent advances in multi-material micro-and nano-manufacturing. *Int. J. Adv. Manuf. Technol.* **2009**, *47*, 161–180. [CrossRef]
10. Leveder, T.; Landis, S.; Davoust, L.; Chaix, N. Flow property measurements for nanoimprint simulation. *Microelectron. Eng.* **2007**, *84*, 928–931. [CrossRef]
11. Reddy, S.; Bonnecaze, R.T. Simulation of fluid flow in the step and flash imprint lithography process. *Microelectron. Eng.* **2005**, *82*, 60–70. [CrossRef]
12. Yanagishita, T.; Kondo, T.; Masuda, H. Preparation of renewable antireflection moth-eye surfaces by nanoimprinting using anodic porous alumina molds. *J. Vac. Sci. Technol. B* **2018**, *36*, 031802. [CrossRef]
13. Sun, J.; Wang, X.; Wu, J.; Jiang, C.; Shen, J.; Cooper, M.A.; Zheng, X.; Liu, Y.; Yang, Z.; Wu, D. Biomimetic moth-eye nanofabrication: Enhanced antireflection with superior self-cleaning characteristic. *Sci. Rep.* **2018**, *8*, 1–10. [CrossRef] [PubMed]
14. Wang, J.; Li, M.; Qiu, J.; Zhou, Y. Filling angle and its effect on filling process in roll-to-roll ultraviolet imprint lithography. *J. Micromech. Microeng.* **2018**, *28*, 035011. [CrossRef]
15. Jain, A.; Spann, A.; Bonnecaze, R.T. Effect of droplet size, droplet placement, and gas dissolution on throughput and defect rate in UV nanoimprint lithography. *J. Vac. Sci. Technol. B* **2017**, *35*, 011602. [CrossRef]
16. Kim, K.-D.; Kwon, H.-J.; Choi, D.-G.; Jeong, J.-H.; Lee, E.-S. Resist flow behavior in ultraviolet nanoimprint lithography as a function of contact angle with stamp and substrate. *Jpn. J. Appl. Phys.* **2008**, *47*, 8648–8651. [CrossRef]
17. Peng, L.; Yi, P.; Wu, H.; Lai, X. Study on bubble defects in roll-to-roll UV imprinting process for micropyramid arrays II: Numerical study. *J. Vac. Sci. Technol. B* **2016**, *34*, 051203. [CrossRef]
18. Wu, H.; Yi, P.; Peng, L.; Lai, X. Study on bubble defects in roll-to-roll UV imprinting process for micropyramid arrays. I. Experiments. *J. Vac. Sci. Technol. B* **2016**, *34*, 021201. [CrossRef]
19. Tian, H.; Shao, J.; Ding, Y.; Li, X.; Li, X. Numerical studies of electrically induced pattern formation by coupling liquid dielectrophoresis and two-phase flow. *Electrophoresis* **2011**, *32*, 2245–2252. [CrossRef] [PubMed]
20. Nagaoka, Y.; Suzuki, R.; Hiroshima, H.; Nishikura, N.; Kawata, H.; Yamazaki, N.; Iwasaki, T.; Hirai, Y. Simulation of resist filling properties under condensable gas ambient in ultraviolet nanoimprint lithography. *Jpn. J. Appl. Phys.* **2012**, *51*, 06FJ07. [CrossRef]
21. Shibata, M.; Horiba, A.; Nagaoka, Y.; Kawata, H.; Yasuda, M.; Hirai, Y. Process-simulation system for UV-nanoimprint lithography. *J. Vac. Sci. Technol. B* **2010**, *28*, C6M108. [CrossRef]
22. Ye, H.; Zhang, Q.; Shen, L.; Li, M. Bubble defect control in low-cost roll-to-roll ultraviolet imprint lithography. *Micro Nano Lett.* **2014**, *9*, 28–30. [CrossRef]
23. Ali, J.; Kim, H.; Cheang, U.K.; Kim, M.J. Micro-PIV measurements of flows induced by rotating microparticles near a boundary. *Microfluid. Nanofluid.* **2016**, *20*, 131. [CrossRef]
24. Cross, B.; Barraud, C.; Picard, C.; Léger, L.; Restagno, F.; Charlaix, É. Wall slip of complex fluids: Interfacial friction versus slip length. *Phys. Rev. Fluids* **2018**, *3*, 062001. [CrossRef]
25. Morihara, D.; Nagaoka, Y.; Hiroshima, H.; Hirai, Y. Numerical study on bubble trapping in UV nanoimprint lithography. *J. Vac. Sci. Technol. B* **2009**, *27*, 2866. [CrossRef]
26. Jain, A.; Bonnecaze, R.T. Fluid management in roll-to-roll nanoimprint lithography. *J. Appl. Phys.* **2013**, *113*, 234511. [CrossRef]
27. Zhang, C.; Zhu, Y.; Yi, P.; Peng, L.; Lai, X. Fabrication of flexible silver nanowire conductive films and transmittance improvement based on moth-eye nanostructure array. *J. Micromech. Microeng.* **2017**, *27*, 075010. [CrossRef]

28. Rashidi, S.; Akar, S.; Bovand, M.; Ellahi, R. Volume of fluid model to simulate the nanofluid flow and entropy generation in a single slope solar still. *Renew. Energy* **2018**, *115*, 400–410. [CrossRef]

29. Wu, W.; Hu, C.; Hu, J.; Yuan, S.; Zhang, R. Jet cooling characteristics for ball bearings using the VOF multiphase model. *Int. J. Therm. Sci.* **2017**, *116*, 150–158. [CrossRef]

30. Hiroshima, H.; Suzuki, K. Study on change in UV nanoimprint pattern by altering shrinkage of UV curable resin. *Jpn. J. Appl. Phys.* **2011**, *50*, 06GK09. [CrossRef]

31. Horiba, A.; Yasuda, M.; Kawata, H.; Okada, M.; Matsui, S.; Hirai, Y. Impact of resist shrinkage and its correction in nanoimprint lithography. *Jpn. J. Appl. Phys.* **2012**, *51*, 06FJ06. [CrossRef]

32. Donggang, Y.; Byung, K. Simulation of the filling process in micro channels for polymeric materials. *J. Micromech. Microeng.* **2002**, *12*, 604.

33. Arabpour, A.; Karimipour, A.; Toghraie, D. The study of heat transfer and laminar flow of kerosene/multi-walled carbon nanotubes (MWCNTs) nanofluid in the microchannel heat sink with slip boundary condition. *J. Therm. Anal. Calorim.* **2017**, *131*, 1553–1566. [CrossRef]

Article

Computational Simulations of Nanoconfined Argon Film through Adsorption–Desorption in a Uniform Slit Pore

Rong-Guang Xu [1],*, Qi Rao [2],*, Yuan Xiang [1], Motong Bian [1] and Yongsheng Leng [1]

[1] Department of Mechanical and Aerospace Engineering, The George Washington University, Washington, DC 20052, USA; xyuan@gwmail.gwu.edu (Y.X.); bianmotong1129@gwmail.gwu.edu (M.B.); leng@gwu.edu (Y.L.)

[2] Energy and Environment Science & Technology, Idaho National Laboratory, Idaho Falls, ID 83415, USA

* Correspondence: xrg117@gwmail.gwu.edu (R.-G.X.); qi.rao@inl.gov (Q.R.); Tel.: +1-410-599-2294 (R.-G.X.); Tel.: +1-202-725-2091 (Q.R.)

Abstract: We performed hybrid grand canonical Monte Carlo/molecular dynamics (GCMC/MD) simulations to investigate the adsorption-desorption isotherms of argon molecules confined between commensurate and incommensurate contacts in nanoscale thickness. The recently proposed mid-density scheme was applied to the obtained hysteresis loops to produce a realistic equilibrium phase of nanoconfined fluids. The appropriate chemical potentials can be determined if the equilibrium structures predicted by GCMC/MD simulations are consistent with those observed in previously developed liquid-vapor molecular dynamics (LVMD) simulations. With the chemical potential as input, the equilibrium structures obtained by GCMC/MD simulations can be used as reasonable initial configurations for future metadynamics free energy calculations.

Keywords: lubrication; nanotribology; phase transition; adsorption; Monte Carlo simulation; molecular dynamics

Citation: Xu, R.-G.; Rao, Q.; Xiang, Y.; Bian, M.; Leng, Y. Computational Simulations of Nanoconfined Argon Film through Adsorption–Desorption in a Uniform Slit Pore. *Coatings* **2021**, *11*, 177. https://doi.org/10.3390/coatings11020177

Received: 14 January 2021
Accepted: 29 January 2021
Published: 2 February 2021

Publisher's Note: MDPI stays neutral with regard to jurisdictional claims in published maps and institutional affiliations.

1. Introduction

The structure and thermodynamic state of a liquid film confined between two solid surfaces in nanoscale thickness and its mechanical response to external loading are very important topics in surface science [1], which has attracted considerable interests and research efforts for decades in many surface force experiments (mainly in surface force apparatus (SFA) or surface force balance (SFB) measurements) [2–5] and molecular simulations [6,7]. The fundamental questions concerning the squeezing and shearing behaviors of nanoconfined fluids are particularly relevant to the fields of nanotribology, biolubrication, coating, and surface engineering and have been a matter of persistent debate despite many new findings and new insights [8–11].

The very low compression rate or lateral sliding speed of mechanical driving springs in an SFA/B instrument largely enables a nanoconfined liquid film in a quasistatic equilibrium state, except in an unstable jump during the layering transition or in a fast slip during a stick-slip sliding. Because of this, we need to first understand the actual equilibrium structure of a nanoconfined liquid film that should correspond to a lowest free energy state. These equilibrium states may not be fully explored by the SFA/B mechanical driving system due to its nonequilibrium nature of driven dynamics during unstable transitions.

Recently, a liquid-vapor molecular dynamics (LVMD) ensemble was used to investigate the force oscillation, phase transition, and shearing behaviors of liquid films confined between two solid walls [12–16]. The main features of the LVMD simulation can be summarized in three aspects: First, it enables the squeeze-out of the confined fluid during normal compression due to the extended side walls along the squeeze-out direction. Second, the lateral pressure of the confined system is comparable to the ambient pressure because of the coexistence of vapor phase around the central liquid droplet. Third, the squeeze-out of the confined fluid is realized by compressing the driving spring, which is connected to the top confining wall. Such a mechanical loading is similar to the situation in SFA/B experiments. Therefore, an LVMD ensemble is particularly suitable in investigating the squeeze-out behavior of confined fluids in surface force experiments. It has also been extended to study the mechanical response of a liquid film confined between AFM tip and substrate [17–19]. A critical issue in the LVMD simulation of a nanoconfined fluid concerns whether the simulation can produce a realistic thermodynamic equilibrium state of the liquid film under a sufficiently low compression rate. In our previous study [14], grand canonical Monte Carlo (GCMC) and LVMD simulations were carried out to study the squeezing and phase transition of simple liquid argon confined between two commensurate solid surfaces. The comparative study revealed that LVMD simulation can capture a major part of the force oscillation profile predicted by the equilibrium GCMC simulation and, at the same time, reveal the non-equilibrium squeeze-out behavior of the confined film. Still, the problem of whether the confined liquid film is in the lowest free energy state remains.

Metadynamics [20,21] proves to be a state-of-the-art computational framework for enhanced sampling coupled with molecular dynamics (MD) simulations in an efficient, flexible, and accurate manner. Its ultimate goal is to simulate rare events and reconstruct the free energy surface (FES) as a function of collective variables (CVs) in physics/biophysics, chemistry, and material science [22,23]. To study phase transitions involving the solid nucleation and/or relative free energy of different phases, structural order parameters (such as Steinhardt parameters [24,25]) can be selected as CVs [26,27]. In this sense, however, LVMD assembly as the simulation setup is not suitable for a metadynamics study, due to the coexistence of multiple phases involved in LVMD simulation. The simulation setup in a uniform slit pore with periodic boundary conditions (PBCs) applied in the lateral directions (i.e., there is no liquid–gas or solid–liquid phase boundary) in our previous GCMC simulations [14] is more suitable for combination with a metadynamics study, since only one homogeneous phase exists in the confined region.

GCMC simulations have been considered a natural way and a standard method to simulate an open system with constant temperature T, volume V, and chemical potential μ (μVT system) imposed by an external reservoir while the number of particles is allowed to vary. Typical examples are clay swelling [28–31] and the quasistatic friction in aqueous film in an SFA experiment [32]. In GCMC simulation, at a given gap width between two confining walls, a hysteresis loop with two distinct branches (adsorption and desorption) is often observed and widely documented, which was not considered in our previous study [14]. How to determine the equilibrium transition within the hysteresis loop is a critical challenge. In this study, a recently proposed method called the mid-density scheme was applied to determine the equilibrium state in a hysteresis loop [33]. Compared with conventional but computationally expensive methods such as thermodynamic integration [34,35] and gauge cell method [36,37], the mid-density scheme is a simplified but useful approach to identify the equilibrium phase transition in a hysteresis loop.

The aim of this paper is to obtain an appropriate value of chemical potential extracted from hybrid grand canonical Monte Carlo/molecular dynamics (GCMC/MD) simulations. The criterion is that it can correctly reproduce the thermodynamic states observed in LVMD simulation, rather than accurately predict the energetically favorable states. In other words, LVMD simulations can serve as a guide in searching the proper value of chemical potential adopted for a GCMC simulation. The equilibrium configuration obtained by this approach can be viewed as a reference and further used as the initial configuration in metadynamics to investigate the free energy profile relative to an order parameter at a given gap distance between confining solid walls.

The rest of the paper is organized as follows. In Section 2, we briefly describe the simulation models and the GCMC simulation techniques. Results are given in Section 3, followed by a summary and discussions in Section 4.

2. Simulation Models and Methods

As illustrated in Figure 1, argon atoms, as a nonpolar model lubricant, are confined between two rigid face-centered cubic (FCC) solid walls. A simple and computationally cheap Lennard–Jones (LJ) atomic pair potential is used to model the interactions between argon molecules according to the literature, i.e., $\sigma = 0.3405$ nm and $\varepsilon = 0.2381$ kcal/mol [38]. Following our previous publications [12,16], two types of contacts are considered: commensurate (Figure 1a,b) and incommensurate (Figure 1c,d). The two FCC crystal walls have the z direction along the [111] direction, while the x and y directions are along the $[1\bar{1}0]$ and $[11\bar{2}]$ directions, respectively. For convenience, the relevant parameters are summarized in Table 1. For both contacts, the wall–fluid interaction between argon and solid wall is the same as the argon–argon interaction (i.e., $\varepsilon_{wf} = \varepsilon_{ff} = \varepsilon$, where w and f stand for "wall" and "fluid," respectively).

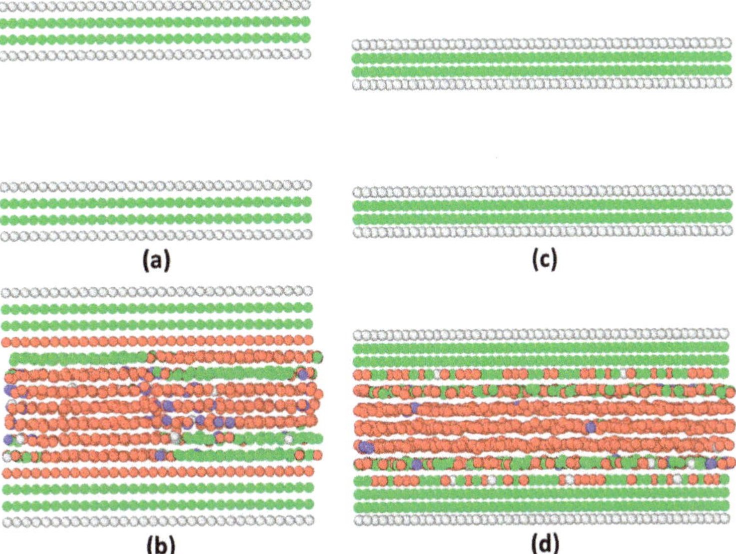

Figure 1. Schematic of the simulation model. Commensurate contact: (**a**) the adsorption branch started with an empty box and (**b**) the desorption branch started with a full box. Incommensurate contact: (**c**) the adsorption branch started with an empty box and (**d**) the desorption branch started with a full box. Atoms in face-centered cubic (FCC), hexagonal close-packed (HCP), body-centered cubic (BCC), and other mixed crystalline structures are in green, red, purple, and white, respectively. The color scheme is consistent throughout this work.

Table 1. The relevant parameters for commensurate and incommensurate contacts used in this study.

	Lattice Constant (nm)	σ_w	σ_{wf}	ε_{wf}	Lateral Dimensions (nm × nm)
Commensurate	0.5405	σ	σ	ε	6.12 × 3.31
Incommensurate	0.408	0.755σ	0.887σ	ε	6.92 × 4.50

Periodic boundary conditions (PBCs) are implemented in the x and y directions to approximate the infinite extent of lateral surfaces. All the simulations are carried out using the Large-scale Atomic/Molecular Massively Parallel Simulator (LAMMPS) package [39]. A standard cutoff distance of 8.5 Å (about 2.5 σ) is used for the LJ interactions, and the time step is 1.0 fs in MD simulations. All the simulations are performed with the Nosé–Hoover thermostat [40,41] at a temperature of 85 K, corresponding to a liquid state of argon which is between the melting point (83.8 K) and the boiling point (87.3 K) of argon.

The local structural ordering of atoms is characterized by the OVITO visualization package [42] with the polyhedral template matching (PTM) method [43]. The PTM method can reliably identify the most common ordered crystalline structures of condensed phases ranging from face-centered cubic (FCC), hexagonal close-packed (HCP), body-centered cubic (BCC), simple cubic (SC), and icosahedral (ICO) to others (unknown coordination structure). It can still work effectively in the presence of strong thermal displacements and strains.

The GCMC simulation conducts exchanges of atoms or molecules with a virtual ideal gas reservoir. Thus, it is considered a natural way to simulate an open system characterized by fixed temperature T, volume V, and chemical potential μ. If all the inserted atoms in the confined region can also be moved using a regular NVT ensemble control (the volume and temperature are held constant), simulations in the grand canonical ensemble (μVT, constant chemical potential, constant volume, and constant temperature) can be performed. In such a hybrid GCMC/MD simulation, the inserted molecules can be well equilibrated.

Our hybrid GCMC/MD simulation attempts 1000 insertion/deletion moves and 1000 translational moves every 100 timesteps. In each simulation, at least 5 million timesteps are performed. The temperature of the imaginary reservoir is set to be equivalent to the target temperature used in NVT ($T = 85$ K) so that the imaginary reservoir is in thermal equilibrium with the simulation cell. When an argon atom is to be inserted, its coordinates are chosen at a random place within the confined region, and new atom velocities are randomly chosen from the specified temperature distribution given by T.

Hybrid GCMC/MD simulations can be initiated with an empty box or a box full of argon molecules, which will give the adsorption and desorption branches respectively. For the former, we continue to increase the chemical potential each time and the resulting configuration of the previous simulation is used as the starting configuration for the subsequent simulation. For the latter, we repeat the same procedure with decreasing chemical potential. In many microporous systems, the adsorption and desorption branches are distinct, exhibiting a hysteresis loop, with the equilibrium phase transition undetermined.

Compared with conventional but computationally expensive methods, the mid-density scheme is a simplified but useful approach to identify the equilibrium phase transition in a hysteresis loop [33]. It has been successfully applied to study argon adsorption in graphite slit pores and in silica cylindrical pores [33,44]. The researchers who proposed this method also demonstrated [45] that its validity for the determination of the equilibrium state is supported by the commonly used canonical ensemble kinetic Monte Carlo (CE-kMC) [46,47] and gauge cell methods [36,37,48].

The detailed procedures are as follows: (1) For each chemical potential, there are two states: low-density state (with average N_{low} argon molecules in the confined region) on the adsorption branch and high-density states (with average N_{high} argon molecules in the confined region) on the desorption branch. (2) The number of particles in the mid-density state is the average of the number of particles in the low-density state and the high-density state, i.e., $N_{mid} = (N_{high} + N_{low})/2$. (3) The mid-density state can be constructed by randomly deleting $(N_{high} - N_{low})/2$ molecules from the high-density state. (4) Run the NVT MD simulation to completely relax the system with N_{mid} particles. (5) As a final step, the hybrid GCMC/MD simulation is performed for a sufficiently long time and the final state obtained is considered to be the stable state. It is summarized in the flowchart shown in Figure 2.

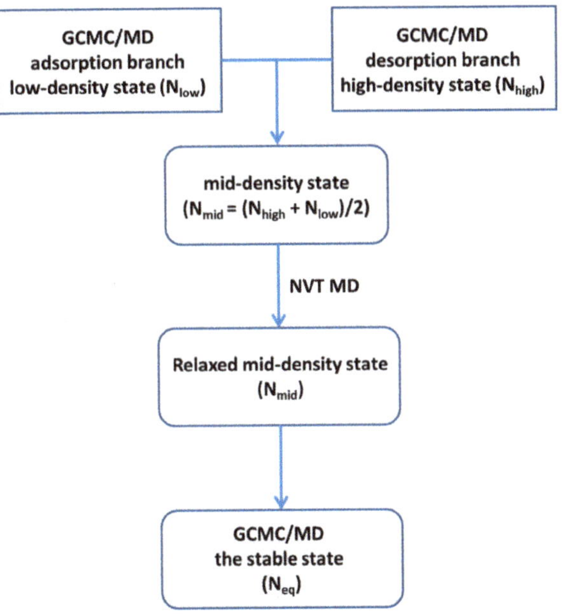

Figure 2. The flowchart of the mid-density scheme applied in this work.

3. Results and Discussion

Our previous LVMD simulations show that the critical layer number, n_c, at which the confinement-induced liquid-like to solid-like phase transition occurs, does not depend on the size of the contact area, but depends on the nature of contact [16]. As to the commensurate contact, this number is $n_c = 7$, while in the case of the incommensurate contact, it is $n_c = 5$, the same critical layer number for mica-cyclohexane-mica contact [15]. Our special interest is in the equilibrium state in both commensurate and incommensurate contacts at n_c and $n_c + 1$ monolayer thickness.

3.1. Commensurate Contact Simulations

For the perfect commensurate contact, hybrid GCMC/MD simulations are performed at $n = 7$ and $n = 8$ monolayer thickness. The adsorption-desorption isotherms are shown in Figure 3. The black and red curves represent adsorption and desorption branches respectively, which form two very broad hysteresis loops. The overall pattern of each adsorption and desorption branch is a three-step shape, representing the gas, liquid and solid phases respectively.

As to $n = 7$ monolayer, the adsorption branch starts with $\mu = -2.40$ kcal/mol, in which the system is in a state of dilute gas phase and the argon atoms are moving close to the confining walls (Figure 4a). As the chemical potential increases, the number of atoms in the gas phase increases and two contact layers are formed (Figure 4b). At $\mu = -2.225$ kcal/mol, the system experiences a gas to liquid phase transition with a sharp increase in the number of particles (Figure 4c). At $\mu = -2.19$ kcal/mol, the system undergoes a liquid to solid phase transition, in which FCC layers and HCP layers alternate to form 7 solid layers (Figure 4d). The desorption branch starts with $\mu = -2.15$ kcal/mol, in which the solidified film is formed in a polycrystalline state that has FCC and HCP crystal structures. The grain boundaries between FCC and HCP crystals are clearly seen in the two outmost contact layers as shown in Figure 4e. The middle three layers are all in HCP structures. Due to existence of grain boundaries and defects, the number of argon particles in such a polycrystalline state is smaller than that in solid state in the adsorption branch in which the packing structure is more compact (Figure 4d). As the chemical potential decreases, the number of atoms in solid state remains constant and the microstructure does not change a lot. At μ = −2.325 kcal/mol, the system experiences a solid (Figure 4f) to liquid phase transition (Figure 4g). At μ = −2.40 kcal/mol, the system goes back to the gas phase (Figure 4h), completing the hysteresis loop.

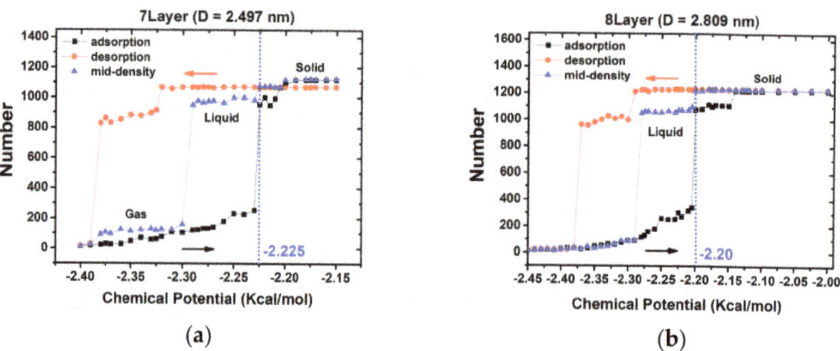

(a)   (b)

Figure 3. Grand canonical Monte Carlo/molecular dynamics (GCMC/MD) simulations of adsorption (black) and desorption (red) isotherms for argon molecules confined between two commensurate contacts at 85 K at (**a**) *n* = 7 monolayer and (**b**) *n* = 8 monolayer thickness. The equilibrium states obtained by the mid-density scheme are shown in blue curves, on which the vertical dashed lines mark the phase boundary between liquid phase and solid phase.

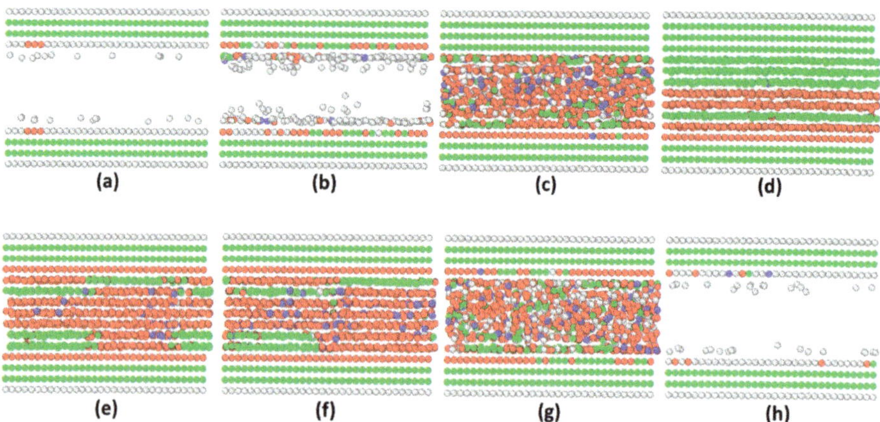

Figure 4. The snapshots of atomic configurations of adsorption-desorption isotherms for argon molecules confined between two commensurate contacts at 85 K for *n* = 7 monolayer thickness. Adsorption: (**a**) μ = −2.40 kcal/mol, (**b**) μ = −2.23 kcal/mol, (**c**) μ = −2.225 kcal/mol, and (**d**) μ = −2.19 kcal/mol. Desorption: (**e**) μ = −2.15 kcal/mol, (**f**) μ = −2.32 kcal/mol, (**g**) μ = −2.325 kcal/mol, and (**h**) μ = −2.40 kcal/mol.

As the next step, the mid-scheme is applied to these adsorption and desorption isotherms to obtain the equilibrium phase as shown in the blue curves in Figure 3a. When μ ≤ −2.30 kcal/mol, the system is in the gas phase and the number particles in the gas phase is notably larger than that in the adsorption branch. When chemical potential is in the range of [−2.29, −2.22] kcal/mol, the equilibrium state is the liquid phase. As can be seen, the solid state exists when chemical potential is larger than −2.225 kcal/mol. Still, two possible solid states are observed: a polycrystalline state similar to that in Figure 4e with smaller chemical potential and 7 solid layers with stacking fault similar to that in Figure 4d corresponding to larger chemical potential.

We repeat the same procedure for the case of $n = 8$ monolayer thickness. The shape of the hysteresis loop is similar to the $n = 7$ monolayer case and, as expected, it is wider. In this case, there is no much difference between the two solid states in adsorption and desorption branches (Figure 5d,e). They are both in the polycrystalline state and the grain boundaries between FCC and HCP crystals are in the two or three outmost contact layers. Based on the mid-density scheme applied to the isotherms, the liquid state is the stable phase when chemical potential is smaller than -2.20 kcal/mol but larger than -2.29 kcal/mol.

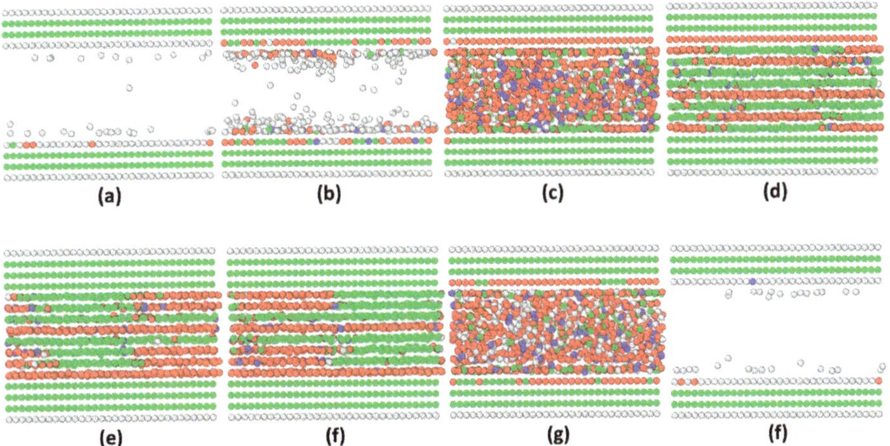

Figure 5. The GCMC snapshots of atomic configurations of adsorption-desorption isotherms for argon molecules confined between two commensurate contacts at 85 K for n = 8 monolayer thickness. Adsorption: (**a**) $\mu = -2.40$ kcal/mol, (**b**) $\mu = -2.21$ kcal/mol, (**c**) $\mu = -2.19$ kcal/mol, and (**d**) $\mu = -2.0$ kcal/mol. Desorption: (**e**) $\mu = -2.20$ kcal/mol, (**f**) $\mu = -2.275$ kcal/mol, (**g**) $\mu = -2.35$ kcal/mol, and (**h**) $\mu = -2.40$ kcal/mol.

Since the critical layer number for the perfect commensurate contact is $n_c = 7$ in LVMD simulations, the confined film is in liquid state at $n = 8$ layers and is in solid state at $n = 7$ layers. Taken together, the appropriate chemical potential should be in the range of $[-2.225, -2.20]$ kcal/mol, which can reproduce the correct equilibrium phase for $n = 7$ and 8 layers observed in LVMD simulations.

In the next step, we select chemical potential $\mu = -2.21$ kcal/mol for the rest of simulations with varying thickness. As can be seen in Table 2, all the equilibrium phases predicted in GCMC simulations (Figure 6) are consistent with the observations of LVMD simulations (see Figure A1), which confirms that $\mu = -2.21$ kcal/mol is an appropriate choice for running GCMC/MD simulations. With this chemical potential, GCMC/MD can produce equilibrium configuration for future simulations coupled with metadynamics.

Table 2. The equilibrium phase in different thicknesses predicted by GCMC/MD simulations with chemical potential $\mu = -2.21$ kcal/mol (Figure 6) and liquid-vapor molecular dynamics (LVMD) simulations (Figure A1).

Thickness	GCMC	LVMD
6.5 Layer + 1.0Å	Liquid	Liquid
6.5 Layer + 0.5Å	Liquid	Liquid
6.5 Layer	Solid	Solid
6 Layer	Solid	Solid
5.5 Layer + 1.0Å	Liquid	Liquid
5.5 Layer + 0.5Å	Liquid	Liquid
5.5 Layer	Solid	Solid
5 Layer	Solid	Solid

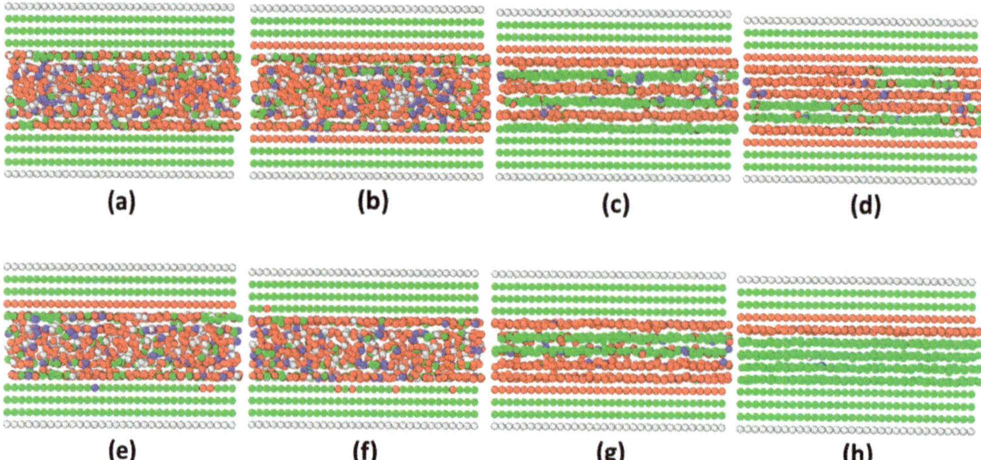

Figure 6. Equilibrium phases of argon molecules confined between two commensurate contacts predicted in GCMC/MD simulations with chemical potential µ = −2.21 kcal/mol for varying thicknesses: (**a**) 6.5 Layer + 1.0Å, (**b**) 6.5 Layer + 0.5Å, (**c**) 6.5 Layer, (**d**) 6 Layer, (**e**) 5.5 Layer + 1.0Å, (**f**) 5.5 Layer + 0.5Å, (**g**) 5.5 Layer, and (**h**) 5 Layer. They are consistent with those observed in LVMD simulations as shown in Figure A1.

3.2. Incommensurate Contact Simulations

We now consider the incommensurate contact between the confining wall and the solidified film. The critical number of monolayers in the film for liquid-like to solid-like phase transition is $n_c = 5$. Following the same procedure in the case of commensurate contact, the hybrid GCMC/MD simulations are performed at $n = 5$ and $n = 6$ monolayer thickness. The adsorption and desorption isotherms are shown in Figure 7. The black and red curves represent adsorption and the desorption branches, respectively. The blue curves show the equilibrium states by applying the mid-density scheme to the adsorption-desorption isotherms. The overall pattern of the hysteresis loop is similar to that in the commensurate case except the desorption branch at $n = 5$ monolayer in which the liquid phase is absent.

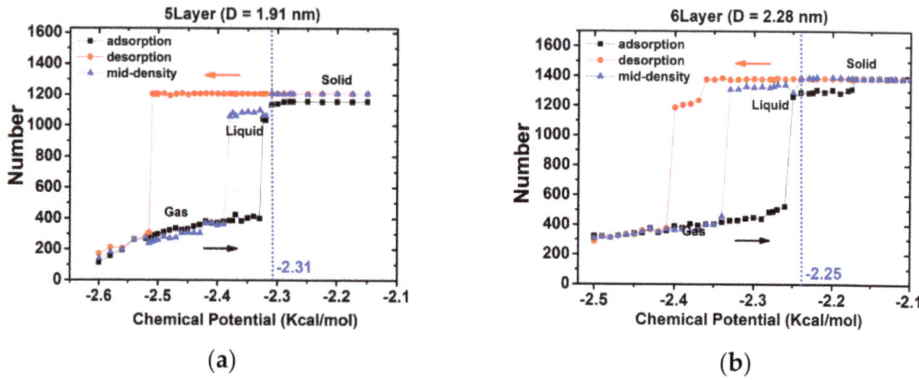

Figure 7. GCMC/MD simulations of adsorption (black) and desorption (red) isotherms for argon molecules confined between two incommensurate contacts at 85 K at (**a**) $n = 5$ monolayer and (**b**) $n = 6$ monolayer thickness. The equilibrium states obtained by the mid-density scheme are shown in blue curves, on which the vertical dashed lines mark the phase boundary between liquid phase and solid phase.

As for the $n = 5$ monolayer, in the adsorption branch, the equilibrium state is the gas phase with chemical potential in the range of $[-2.60, -2.33]$ kcal/mol (Figure 8a). Liquid state (Figure 8b) can only exist in a very narrow range (from -2.325 kcal/mol to -2.320 kcal/mol), which is followed by solid phase when chemical potential $\mu \geq -2.31$. It should be noted that the normal of the (111) close packed plane of the solid structure is nearly along the diagonal direction in the lateral x–y plane (Figure 8c). The desorption branch starts with $\mu = -2.15$ kcal/mol, in which the solidified film is formed. Compared with the solidified structure in the adsorption branch (Figure 8c), the lattice orientation is different and its close packed normal is off the diagonal of the lateral x–y plane (Figure 8d). The slightly different configurations can explain the difference in the number of particles in solidified structures (Figure 8c,d). As the chemical potential decreases, the number of atoms in solid state remains constant and the microstructure does not change a lot (Figure 8e). At $\mu = -2.515$ kcal/mol, the system experiences a solid (Figure 8e) to gas phase transition (Figure 8f), completing a hysteresis loop. In this case, no liquid phase is observed. In the next step, the mid-scheme is applied to these adsorption and desorption isotherms to obtain the equilibrium phase as shown in the blue curves in Figure 7a. It can be seen that the equilibrium phase is the solid state when chemical potential is larger than -2.31 kcal/mol. For the case of n = 6 monolayer, even though the lattice orientations of the solidified structures in the adsorption and desorption branches are different (Figure 9c,d), there is no distinct difference in the number of particles in the solid structures. As can be seen, the liquid state is the stable phase when chemical potential is in the range of $[-2.33, -2.25]$ kcal/mol. Considering the critical layer number for the incommensurate contact being $n_c = 5$ in LVMD simulations, the appropriate chemical potential should be in the range of $[-2.31, -2.25]$ kcal/mol, which can reproduce the correct equilibrium phase for $n = 5$ (in solid state) and $n = 6$ (in liquid state) monolayer observed in LVMD simulations. In the following, chemical potential $\mu = -2.28$ kcal/mol is adopted for the rest of the simulations with varying thicknesses as shown in Table 3. All the equilibrium phases predicted in GCMC simulations are consistent with the observations of LVMD simulations, confirming that $\mu = -2.28$ kcal/mol is an appropriate choice.

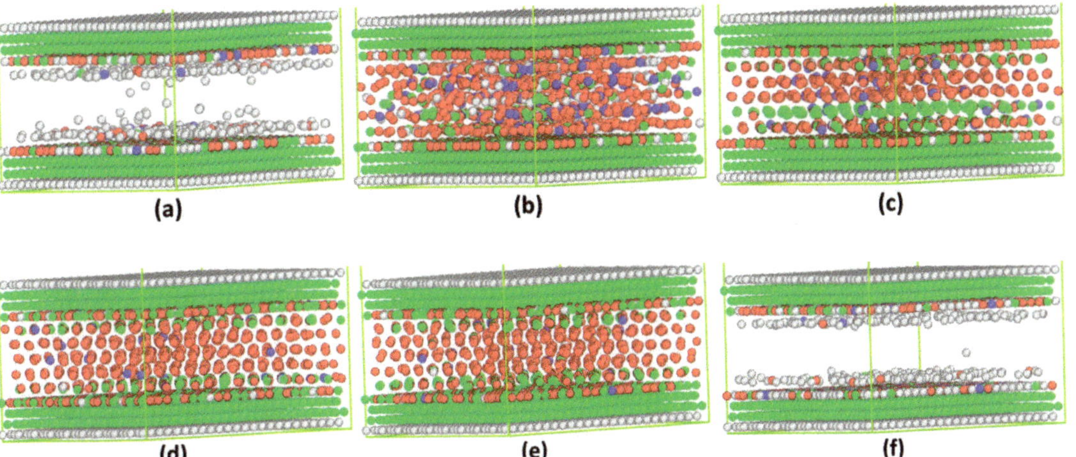

Figure 8. The GCMC snapshots of the atomic configurations of the adsorption-desorption isotherms for argon molecules confined between two incommensurate contacts at 85 K for $n = 5$ monolayer thickness. Adsorption: (**a**) $\mu = -2.33$ kcal/mol, (**b**) $\mu = -2.32$ kcal/mol, and (**c**) $\mu = -2.15$ kcal/mol. Desorption: (**d**) $\mu = -2.15$ kcal/mol, (**e**) $\mu = -2.51$ kcal/mol, and (**f**) $\mu = -2.515$ kcal/mol.

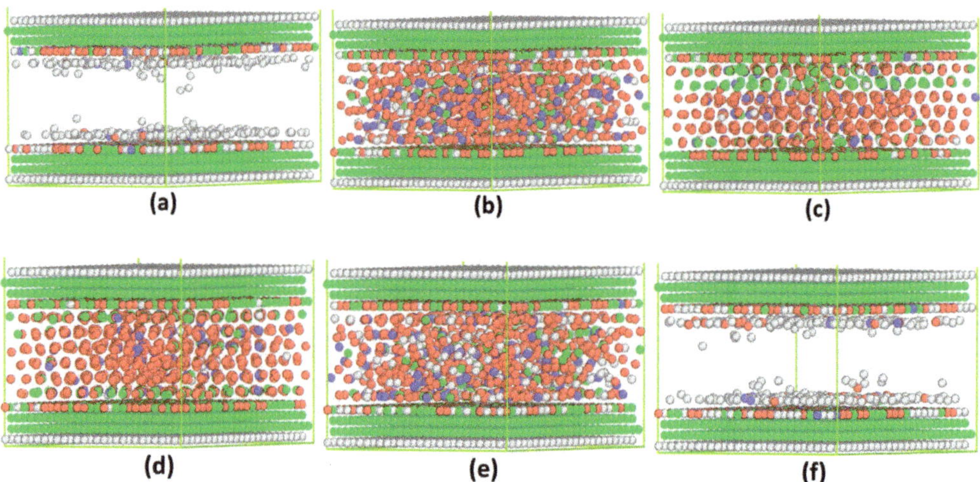

Figure 9. The GCMC snapshots of atomic configurations of adsorption-desorption isotherms for argon molecules confined between two incommensurate contacts at 85 K for n = 6 monolayer thickness. Adsorption: (**a**) $\mu = -2.26$ kcal/mol, (**b**) $\mu = -2.25$ kcal/mol, and (**c**) $\mu = -2.15$ kcal/mol. Desorption: (**d**) $\mu = -2.15$ kcal/mol, (**e**) $\mu = -2.37$ kcal/mol, and (**f**) $\mu = -2.41$ kcal/mol.

Table 3. The equilibrium phase in different thicknesses predicted by GCMC simulations with chemical potential $\mu = -2.28$ kcal/mol (Figure 10) and LVMD simulations (Figure A2).

Thickness	GCMC	LVMD
18.5 Å	Solid	Solid
17.5 Å	Liquid	Liquid
17.0 Å	Liquid	Liquid
16.0 Å	Solid	Solid
15.0 Å	Solid	Solid
14.0 Å	Liquid	Liquid
13.0 Å	Solid	Solid
12.5 Å	Solid	Solid

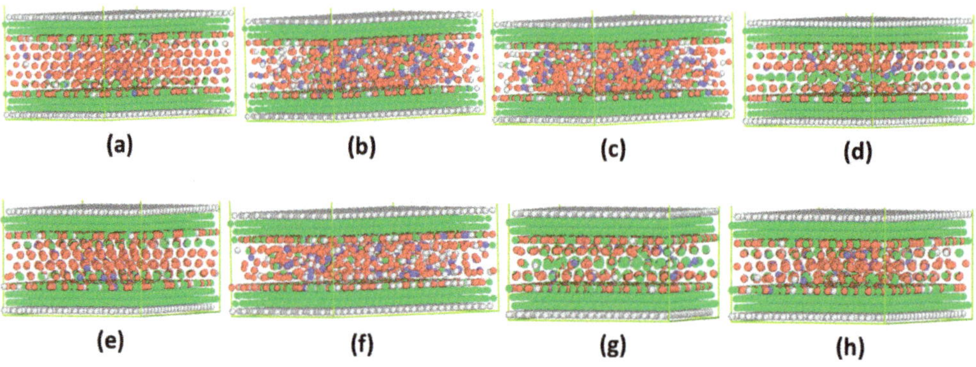

Figure 10. Equilibrium phases of argon molecules confined between two incommensurate contacts predicted in GCMC/MD simulations with chemical potential $\mu = -2.28$ kcal/mol for varying thicknesses: (**a**) 18.5 Å, (**b**) 17.5 Å, (**c**) 17.0 Å, (**d**) 16.0 Å, (**e**) 15.0 Å, (**f**) 14.0 Å, (**g**) 13.0 Å, and (**h**) 12.5 Å. They are consistent with those observed in LVMD simulations as shown in Figure A2.

4. Summary and Further Discussion

It has been discussed that LVMD ensemble is not suitable for metadynamics simulations, because of the coexistence of multi-phases. Instead, simple geometry with slit pores can be a proper simulation setup. In this work, hybrid GCMC/MD simulations are carried out to investigate the adsorption and desorption isotherms of argon molecules confined between commensurate and incommensurate contacts in nanoscale thickness. By applying the mid-density scheme to the hysteresis loops, the equilibrium structures associated with n_c and $n_c + 1$ layers can be obtained. The appropriate chemical potentials can be determined if the equilibrium structures predicted by GCMC/MD simulations are consistent with those observed in LVMD simulations. Such equilibrium structures at varying thicknesses obtained using GCMC/MD simulations can be used as reasonable initial configurations for further metadynamics studies to investigate the free energy profile as a function of structural order parameters.

The shift of the chemical potential due to the presence of nanoconfinement makes the selection of chemical potential μ difficult, and in many situations the choice is merely based on the bulk chemical potential. In our previous study [14], we discussed that the bulk chemical potential μ for argon at around 85 K obtained either from Widom's insertion method in MC calculations or from the integral equation theory [49] cannot be used in GCMC simulation. In this study, we find that the appropriate chemical potential is in the range of $[-2.225, 2.20]$ kcal/mol for the commensurate contact and $[-2.31, -2.25]$ kcal/mol for the incommensurate contact. There is even no overlap between them, which means that no unique chemical potential can be adopted for both cases. Given the fact that confinement usually induces a shift in chemical potential when considering the equilibrium between the slit pore and bulk fluid, we expect that the shift due to the different nature of contact should also be different.

It should be noted that in previous works, researchers mainly focus on the capillary condensation (the gas-liquid transition) in the pore, in which only two phases, gas and liquid, are involved in the GCMC simulations. When the confined space comes down to the boundary lubrication (BL) regime, three phases come into play, which makes the hysteresis loop more complicated and compounds the problems of determining the equilibrium state. As a first attempt, we apply the mid-density scheme to such a hysteresis loop. There is still some uncertainty about whether the mid-density scheme can be applied to the solid phase. However, our purpose is not to construct equilibrium phase transition in a very accurate and rigorous manner, but to construct a reasonable initial configuration at a given gap distance by GCMC/MD simulations to further investigate the free energy profile vs. structural order parameters. To accurately identify the location of the phase equilibrium, the gauge cell MC method [36,37] needs to be applied in future work.

Author Contributions: Conceptualization, R.-G.X., Q.R., and Y.L.; methodology, R.-G.X., Q.R., Y.X., and M.B.; software, R.-G.X. and M.B.; validation, R.-G.X., Q.R., and Y.X.; formal analysis, R.-G.X. and Q.R.; investigation, R.-G.X.; resources, R.-G.X.; data curation, R.-G.X.; writing—original draft preparation, R.-G.X.; writing—review and editing, R.-G.X., Q.R., and Y.L.; visualization, R.-G.X.; supervision, R.-G.X.; project administration, R.-G.X. and Y.L.; funding acquisition, Y.L. All authors have read and agreed to the published version of the manuscript.

Funding: This work is supported by the American Chemical Society Petroleum Research Fund (ACS PRF), the National Science Foundation (NSF 1149704) and the National Energy Research Scientific Computing Center (NERSC).

Data Availability Statement: Data are contained within the article.

Conflicts of Interest: The authors declare no conflict of interest.

Appendix A. Equilibrium Phase in LVMD Simulations

The equilibrium LVMD simulations of argon molecules confined between commensurate and incommensurate contacts are performed at a certain wall distance. It is found that the equilibrium configurations observed in LVMD simulations are consistent with those observed in GCMC/MD simulations. It should be noted that in this work, the computational cost of both LVMD and GCMC/MD simulations are within the same order of magnitude (every single LVMD or GCMC/MD simulation at each gap thickness is about a few hours long).

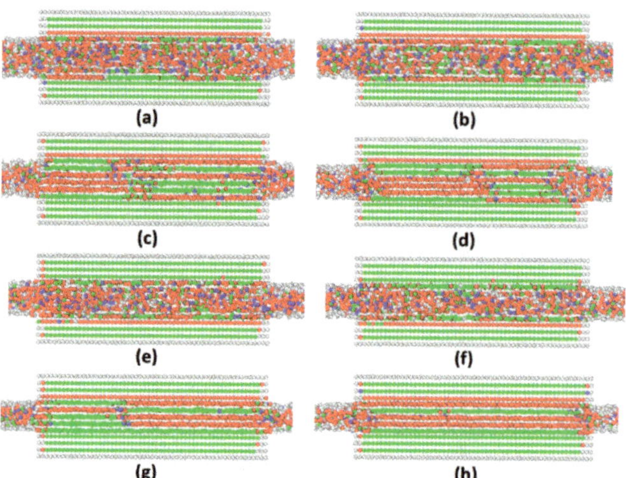

Figure A1. The equilibrium configurations of argon molecules confined between two commensurate contacts observed in LVMD simulations with varying thicknesses: (**a**) 6.5 Layer + 1.0Å, (**b**) 6.5 Layer + 0.5Å, (**c**) 6.5 Layer, (**d**) 6 Layer, (**e**) 5.5 Layer + 1.0Å, (**f**) 5.5 Layer + 0.5Å, (**g**) 5.5 Layer, and (**h**) 5 Layer. Only central walls are shown. They are consistent with those observed in GCMC/MD simulations with chemical potential $\mu = -2.21$ kcal/mol, as shown in Figure 6.

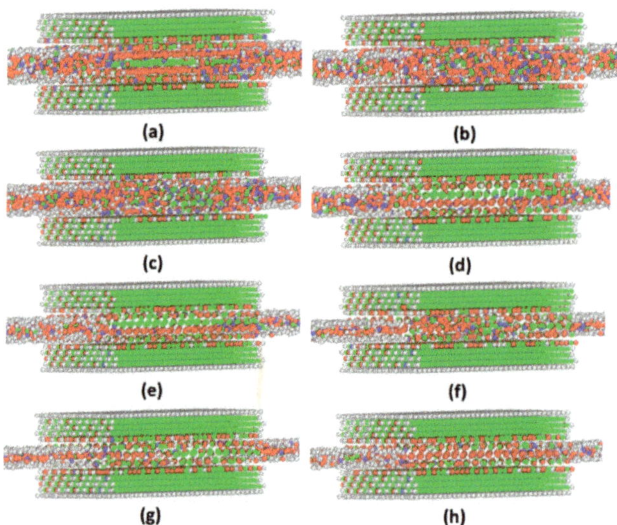

Figure A2. The equilibrium configurations of argon molecules confined between two incommensurate contacts observed in LVMD simulations with varying thicknesses: (**a**) 18.5 Å, (**b**) 17.5 Å, (**c**) 17.0 Å, (**d**) 16.0 Å, (**e**) 15.0 Å, (**f**) 14.0 Å, (**g**) 13.0 Å, and (**h**) 12.5 Å. Only central walls are shown. They are consistent with those observed in GCMC/MD simulations with chemical potential $\mu = -2.28$ kcal/mol as shown in Figure 10.

References

1. Israelachvili, J.N. *Intermolecular and Surface Forces*, 3rd ed.; Elsevier, Academic Press, Inc.: San Diego, CA, USA, 2011.
2. Klein, J.; Kumacheva, E. Confinement-induced phase transitions in simple liquids. *Science* **1995**, *269*, 816–819. [CrossRef] [PubMed]
3. Klein, J.; Kumacheva, E. Simple liquids confined to molecularly thin layers. I. Confinement-induced liquid-to-solid phase transitions. *J. Chem. Phys.* **1998**, *108*, 6996–7009. [CrossRef]
4. Kumacheva, E.; Klein, J. Simple liquids confined to molecularly thin layers. II. Shear and frictional behavior of solidified films. *J. Chem. Phys.* **1998**, *108*, 7010–7022. [CrossRef]
5. Demirel, A.L.; Granick, S. Glasslike transition of a confined simple fluid. *Phys. Rev. Lett.* **1996**, *77*, 2261–2264. [CrossRef] [PubMed]
6. Thompson, P.A.; Robbins, M.O. Origin of stick-slip motion in boundary lubrication. *Science* **1990**, *250*, 792–794. [CrossRef]
7. Docherty, H.; Cummings, P.T. Direct evidence for fluid–solid transition of nanoconfined fluids. *Soft Matter* **2010**, *6*, 1640–1643. [CrossRef]
8. Rosenhek-Goldian, I.; Kampf, N.; Yeredor, A.; Klein, J. On the question of whether lubricants fluidize in stick–slip friction. *Proc. Natl. Acad. Sci. USA* **2015**, *112*, 7117–7122. [CrossRef]
9. Smith, A.M.; Hallett, J.E.; Perkin, S. Solidification and superlubricity with molecular alkane films. *Proc. Natl. Acad. Sci. USA* **2019**, *116*, 25418–25423. [CrossRef]
10. Gooneie, A.; Balmer, T.E.; Heuberger, M. Seclusion of molecular layers in a confined simple liquid. *Phys. Rev. Res.* **2020**, *2*, 022026. [CrossRef]
11. Gao, H.; Müser, M.H. Why liquids can appear to solidify during squeeze-out–Even when they don't. *J. Colloid Interface Sci.* **2020**, *562*, 273–278. [CrossRef]
12. Lei, Y.; Leng, Y. Force oscillation and phase transition of simple fluids under confinement. *Phys. Rev. E* **2010**, *82*, 040501. [CrossRef] [PubMed]
13. Lei, Y.; Leng, Y. Stick-slip friction and energy dissipation in boundary lubrication. *Phys. Rev. Lett.* **2011**, *107*, 147801. [CrossRef] [PubMed]
14. Leng, Y.; Xiang, Y.; Lei, Y.; Rao, Q. A comparative study by the grand canonical Monte Carlo and molecular dynamics simulations on the squeezing behavior of nanometers confined liquid films. *J. Chem. Phys.* **2013**, *139*, 074704. [CrossRef] [PubMed]
15. Xu, R.-G.; Leng, Y. Squeezing and stick–slip friction behaviors of lubricants in boundary lubrication. *Proc. Natl. Acad. Sci. USA* **2018**, *115*, 6560–6565. [CrossRef]
16. Xu, R.-G.; Xiang, Y.; Papanikolaou, S.; Leng, Y. On the shear dilation of polycrystalline lubricant films in boundary lubricated contacts. *J. Chem. Phys.* **2020**, *152*, 104708. [CrossRef]
17. Xu, R.-G.; Leng, Y. Solvation force simulations in atomic force microscopy. *J. Chem. Phys.* **2014**, *140*, 214702. [CrossRef]
18. Xu, R.-G.; Leng, Y. Contact stiffness and damping of liquid films in dynamic atomic force microscope. *J. Chem. Phys.* **2016**, *144*, 154702. [CrossRef]
19. Xu, R.-G.; Xiang, Y.; Leng, Y. Computational simulations of solvation force and squeezing out of dodecane chain molecules in an atomic force microscope. *J. Chem. Phys.* **2017**, *147*, 054705. [CrossRef]
20. Laio, A.; Parrinello, M. Escaping free-energy minima. *Proc. Natl. Acad. Sci. USA* **2002**, *99*, 12562–12566. [CrossRef]
21. Barducci, A.; Bussi, G.; Parrinello, M. Well-tempered metadynamics: A smoothly converging and tunable free-energy method. *Phys. Rev. Lett.* **2008**, *100*, 020603. [CrossRef]
22. Barducci, A.; Bonomi, M.; Parrinello, M. Metadynamics. *Wiley Interdiscip. Rev. Comput. Mol. Sci.* **2011**, *1*, 826–843. [CrossRef]
23. Laio, A.; Gervasio, F.L. Metadynamics: A method to simulate rare events and reconstruct the free energy in biophysics, chemistry and material science. *Rep. Prog. Phys.* **2008**, *71*, 126601. [CrossRef]
24. Steinhardt, P.J.; Nelson, D.R.; Ronchetti, M. Bond-orientational order in liquids and glasses. *Phys. Rev. B* **1983**, *28*, 784–805. [CrossRef]
25. Lechner, W.; Dellago, C. Accurate determination of crystal structures based on averaged local bond order parameters. *J. Chem. Phys.* **2008**, *129*, 114707. [CrossRef]
26. Tribello, G.A.; Giberti, F.; Sosso, G.C.; Salvalaglio, M.; Parrinello, M. Analyzing and driving cluster formation in atomistic simulations. *J. Chem. Theory Comput.* **2017**, *13*, 1317–1327. [CrossRef]
27. Giberti, F.; Salvalaglio, M.; Parrinello, M. Metadynamics studies of crystal nucleation. *IUCrJ* **2015**, *2*, 256–266. [CrossRef]
28. Tambach, T.J.; Bolhuis, P.G.; Hensen, E.J.; Smit, B. Hysteresis in clay swelling induced by hydrogen bonding: Accurate prediction of swelling states. *Langmuir* **2006**, *22*, 1223–1234. [CrossRef]
29. Rao, Q.; Xiang, Y.; Leng, Y. Molecular simulations on the structure and dynamics of water–methane fluids between Na-montmorillonite clay surfaces at elevated temperature and pressure. *J. Phys. Chem. C* **2013**, *117*, 14061–14069. [CrossRef]
30. Rao, Q.; Leng, Y. Methane aqueous fluids in montmorillonite clay interlayer under near-surface geological conditions: A grand canonical Monte Carlo and molecular dynamics simulation study. *J. Phys. Chem. B* **2014**, *118*, 10956–10965. [CrossRef]
31. Rao, Q.; Leng, Y. Molecular understanding of CO_2 and H_2O in a montmorillonite clay interlayer under CO_2 geological sequestration conditions. *J. Phys. Chem. C* **2016**, *120*, 2642–2654. [CrossRef]
32. Fedyanin, I.; Pertsin, A.; Grunze, M. Quasistatic computer simulations of shear behavior of water nanoconfined between mica surfaces. *J. Chem. Phys.* **2011**, *135*, 174704. [CrossRef] [PubMed]

33. Liu, Z.; Herrera, L.; Nguyen, V.T.; Do, D.; Nicholson, D. A Monte Carlo scheme based on mid-density in a hysteresis loop to determine equilibrium phase transition. *Mol. Simul.* **2011**, *37*, 932–939. [CrossRef]
34. Peterson, B.K.; Gubbins, K.E. Phase transitions in a cylindrical pore: Grand canonical Monte Carlo, mean field theory and the Kelvin equation. *Mol. Phys.* **1987**, *62*, 215–226. [CrossRef]
35. Peterson, B.K.; Gubbins, K.E.; Heffelfinger, G.S.; Marini Bettolo Marconi, U.; van Swol, F. Lennard-Jones fluids in cylindrical pores: Nonlocal theory and computer simulation. *J. Chem. Phys.* **1988**, *88*, 6487–6500. [CrossRef]
36. Neimark, A.V.; Vishnyakov, A. Gauge cell method for simulation studies of phase transitions in confined systems. *Phys. Rev. E* **2000**, *62*, 4611–4622. [CrossRef]
37. Vishnyakov, A.; Neimark, A.V. Studies of Liquid—Vapor Equilibria, Criticality, and Spinodal Transitions in Nanopores by the Gauge Cell Monte Carlo Simulation Method. *J. Phys. Chem. B* **2001**, *105*, 7009–7020. [CrossRef]
38. Allen, M.P.; Tildesley, D.J. *Computer Simulation of Liquids*; Oxford University Press: Oxford, UK, 2017.
39. Plimpton, S. Fast parallel algorithms for short-range molecular dynamics. *J. Comput. Phys.* **1995**, *117*, 1–19.
40. Hoover, W.G. Canonical dynamics: Equilibrium phase-space distributions. *Phys. Rev. A* **1985**, *31*, 1695–1697. [CrossRef]
41. Nosé, S. A unified formulation of the constant temperature molecular dynamics methods. *J. Chem. Phys.* **1984**, *81*, 511–519. [CrossRef]
42. Stukowski, A. Visualization and analysis of atomistic simulation data with OVITO–the open visualization tool. *Model. Simul. Mater. Sci. Eng.* **2009**, *18*, 015012. [CrossRef]
43. Larsen, P.M.; Schmidt, S.; Schiøtz, J. Robust structural identification via polyhedral template matching. *Model. Simul. Mater. Sci. Eng.* **2016**, *24*, 055007. [CrossRef]
44. Liu, Z.; Do, D.D.; Nicholson, D. A thermodynamic study of the mid-density scheme to determine the equilibrium phase transition in cylindrical pores. *Mol. Simul.* **2012**, *38*, 189–199. [CrossRef]
45. Nguyen, V.T.; Do, D.; Nicholson, D. Reconciliation of different simulation methods in the determination of the equilibrium branch for adsorption in pores. *Mol. Simul.* **2014**, *40*, 713–720. [CrossRef]
46. Ustinov, E.A.; Do, D. Application of kinetic Monte Carlo method to equilibrium systems: Vapour–liquid equilibria. *J. Colloid Interface Sci.* **2012**, *366*, 216–223. [CrossRef]
47. Fan, C.; Do, D.; Nicholson, D.; Ustinov, E. Chemical potential, Helmholtz free energy and entropy of argon with kinetic Monte Carlo simulation. *Mol. Phys.* **2014**, *112*, 60–73. [CrossRef]
48. Nguyen, V.T.; Do, D.; Nicholson, D. Monte Carlo simulation of the gas-phase volumetric adsorption system: Effects of dosing volume size, incremental dosing amount, pore shape and size, and temperature. *J. Phys. Chem. B* **2011**, *115*, 7862–7871. [CrossRef]
49. Kast, S.M. Combinations of simulation and integral equation theory. *Phys. Chem. Chem. Phys.* **2001**, *3*, 5087–5092. [CrossRef]

MDPI

St. Alban-Anlage 66

4052 Basel

Switzerland

Tel. +41 61 683 77 34

Fax +41 61 302 89 18

www.mdpi.com

Coatings Editorial Office

E-mail: coatings@mdpi.com

www.mdpi.com/journal/coatings